McGraw-Hill My Math

Welcome to *My Math* – your very own math book! You can write in it – in fact, you are encouraged to write, draw, circle, explain, and color as you explore the exciting world of mathematics. Let's get started. Grab a pencil and finish each sentence.

My name is _____.

My favorite color is _____.

My favorite hobby or sport is _____.

My favorite TV program or video game is

_____.

My favorite class is _____.

Mc Graw Hill Education

mhmymath.com

STEM McGraw-Hill is committed to providing
instructional materials in Science, Technology, Engineering,
and Mathematics (STEM) that give all students a solid
foundation, one that prepares them for college and careers
in the 21st century.

Send all inquiries to:
McGraw-Hill Education
8787 Orion Place
Columbus, OH 43240

ISBN: 978-0-07-678996-2 (**Volume 1**)
MHID: 0-07-678996-9

Printed in the United States of America.

6 7 8 9 LWI 23 22 21 20

Understanding by Design® is a registered trademark of the Association for Supervision and
Curriculum Development ("ASCD").

McGraw-Hill

My Math

Grade 1 • Volume 1

Authors:
Carter • Cuevas • Day • Malloy
Altieri • Balka • Gonsalves • Grace • Krulik • Molix-Bailey
Moseley • Mowry • Myren • Price • Reynosa • Santa Cruz
Silbey • Vielhaber

Mc
Graw
Hill
Education

GO digital ▶ connectED.mcgraw-hill.com

▶ Log In

1 Go to
connectED.mcgraw-hill.com.

2 Log in using your username and password.

3 Click on the Student Edition icon to open the Student Center.

▶ Go to the Student Center

4 Click on Menu, then click on the **Resources** tab to see all of your online resources arranged by chapter and lesson.

5 Click on the **eToolkit** in the Lesson Resources section to open a library of eTools and virtual manipulatives.

6 Look here to find any assignments or messages from your teacher.

7 Click on the **eBook** to open your online Student Edition.

▶ Explore the eBook!

8 Click the **speaker icon** at the top of the eBook pages to hear the page read aloud to you.

More resources can be found by clicking the icons at the bottom of the eBook pages.

 Practice and review your Vocabulary.

 Animations and videos allow you to explore mathematical topics.

 Explore concepts with eTools and virtual manipulatives.

 eHelp helps you complete your homework.

 Explore these fun digital activities to practice what you learned in the classroom.

 Worksheets are PDFs for Math at Home, Problem of the Day, and Fluency Practice.

Contents in Brief
Organized by Domain

Operations and Algebraic Thinking

Chapter 1   Addition Concepts
Chapter 2   Subtraction Concepts
Chapter 3   Addition Strategies to 20
Chapter 4   Subtraction Strategies to 20

Number and Operations in Base Ten

Chapter 5   Place Value
Chapter 6   Two-Digit Addition and Subtraction

Measurement and Data

Chapter 7   Organize and Use Graphs
Chapter 8   Measurement and Time

Geometry

Chapter 9   Two-Dimensional Shapes and Equal Shares
Chapter 10 Three-Dimensional Shapes

Processes & Practices → Woven Throughout

connectED.mcgraw-hill.com

Chapter

Addition Concepts

Getting Started

Chapter 1 Project 2
Am I Ready? 3
My Math Words 4
My Vocabulary Cards 5
My Foldable **FOLDABLES** 9

Lessons and Homework

Lesson 1   Addition Stories 11
Lesson 2   Model Addition 17
Lesson 3   Addition Number Sentences . . . 23
Lesson 4   Add 0 29
Check My Progress 35
Lesson 5   Vertical Addition 37
Lesson 6   Problem-Solving Strategy:
            Write a Number Sentence 43
Lesson 7   Ways to Make 4 and 5 49
Lesson 8   Ways to Make 6 and 7 55
Lesson 9   Ways to Make 8 61
Check My Progress 67
Lesson 10   Ways to Make 9 69
Lesson 11   Ways to Make 10 75
Lesson 12   Find Missing Parts of 10 81
Lesson 13   True and False Statements 87

Wrap Up

Fluency Practice 93
My Review 95
Reflect 98
Performance Task 98PT1

Let's explore more online!

connectED.mcgraw-hill.com

Chapter 2 Subtraction Concepts

Getting Started

Chapter 2 Project 100
Am I Ready? 101
My Math Words 102
My Vocabulary Cards 103
My Foldable **FOLDABLES** 107

Lessons and Homework

Lesson 1   Subtraction Stories 109
Lesson 2   Model Subtraction 115
Lesson 3   Subtraction Number Sentences 121
Lesson 4   Subtract 0 and All 127
Lesson 5   Vertical Subtraction 133
Check My Progress 139
Lesson 6   Problem-Solving Strategy: Draw a Diagram . . . 141
Lesson 7   Compare Groups 147
Lesson 8   Subtract from 4 and 5 153
Lesson 9   Subtract from 6 and 7 159
Check My Progress 165
Lesson 10   Subtract from 8 167
Lesson 11   Subtract from 9 173
Lesson 12   Subtract from 10 179
Lesson 13   Relate Addition and Subtraction 185
Lesson 14   True and False Statements 191

Wrap Up

Fluency Practice 197
My Review 199
Reflect 202
Performance Task 202PT1

Your safari adventure starts online!

connectED.mcgraw-hill.com

Chapter

 3 Addition Strategies to 20

Getting Started

Chapter 3 Project **204**
Am I Ready? **205**
My Math Words **206**
My Vocabulary Cards **207**
My Foldable FOLDABLES **209**

Lessons and Homework

Lesson 1   Count On 1, 2, or 3 . . . . **211**
Lesson 2   Count On Using Pennies **217**
Lesson 3   Use a Number Line to Add **223**
Lesson 4   Use Doubles to Add **229**
Lesson 5   Use Near Doubles to Add **235**
Check My Progress **241**
Lesson 6   Problem-Solving Strategy: Act It Out **243**
Lesson 7   Make 10 to Add **249**
Lesson 8   Add in Any Order **255**
Lesson 9   Add Three Numbers **261**

Wrap Up

Fluency Practice **267**
My Review **269**
Reflect **272**
Performance Task **272PT1**

Look for this!
Click online and you can watch videos that will help you learn the lessons.

Watch

Chapter

Subtraction Strategies to 20

Getting Started

Chapter 4 Project **274**
Am I Ready? **275**
My Math Words **276**
My Vocabulary Cards **277**
My Foldable FOLDABLES **279**

Lessons and Homework

Lesson 1   Count Back 1, 2, or 3 . . . . **281**
Lesson 2   Use a Number Line to Subtract **287**
Lesson 3   Use Doubles to Subtract **293**
Lesson 4   Problem-Solving Strategy:
           Write a Number Sentence **299**
Check My Progress **305**
Lesson 5   Make 10 to Subtract **307**
Lesson 6   Use Related Facts to Add and Subtract **313**
Lesson 7   Fact Families **319**
Lesson 8   Missing Addends **325**

Wrap Up

Fluency Practice **331**
My Review **333**
Reflect **336**
Performance Task **336PT1**

connectED.mcgraw-hill.com

Chapter
5 Place Value

Getting Started

Chapter 5 Project **338**
Am I Ready? 339
My Math Words 340
My Vocabulary Cards 341
My Foldable **FOLDABLES** 345

Lessons and Homework

Lesson 1   Numbers 11 to 19 . . . . **347**
Lesson 2   Tens **353**
Lesson 3   Count by Tens Using Dimes **359**
Lesson 4   Ten and Some More **365**
Lesson 5   Tens and Ones **371**
Check My Progress **377**
Lesson 6   Problem-Solving Strategy: Make a Table **379**
Lesson 7   Numbers to 100 . . . . **385**
Lesson 8   Ten More, Ten Less **391**
Lesson 9   Count by Fives Using Nickels **397**
Lesson 10   Use Models to Compare Numbers **403**
Lesson 11   Use Symbols to Compare Numbers **409**
Check My Progress **415**
Lesson 12   Numbers to 120 . . . . **417**
Lesson 13   Count to 120 . . . . **423**
Lesson 14   Read and Write Numbers to 120 . . . . . **429**

Wrap Up

My Review **435**
Reflect **438**
Performance Task **438PT1**

Chapter 6
Two-Digit Addition and Subtraction

Number and Operations in Base Ten

ESSENTIAL QUESTION
How can I add and subtract two-digit numbers?

Getting Started

Chapter 6 Project **440**
Am I Ready? **441**
My Math Words **442**
My Vocabulary Cards **443**
My Foldable **FOLDABLES** **445**

Lessons and Homework

Lesson 1   Add Tens **447**
Lesson 2   Count On Tens and Ones **453**
Lesson 3   Add Tens and Ones **459**
Lesson 4   Problem-Solving Strategy:
            Guess, Check, and Revise **465**
Lesson 5   Add Tens and Ones with Regrouping **471**
Check My Progress **477**
Lesson 6   Subtract Tens **479**
Lesson 7   Count Back by 10s **485**
Lesson 8   Relate Addition and Subtraction of Tens **491**

Wrap Up

My Review **497**
Reflect **500**
Performance Task **500PT1**

You can find fun activities online!

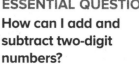

connectED.mcgraw-hill.com

Chapter

7 Organize and Use Graphs

Measurement and Data

ESSENTIAL QUESTION
How do I make and read graphs?

Getting Started

Chapter 7 Project **502**
Am I Ready? **503**
My Math Words **504**
My Vocabulary Cards **505**
My Foldable FOLDABLES **507**

Lessons and Homework

Lesson 1   Tally Charts **509**
Lesson 2   Problem-Solving Strategy: Make a Table **515**
Lesson 3   Make Picture Graphs **521**
Lesson 4   Read Picture Graphs **527**
Check My Progress **533**
Lesson 5   Make Bar Graphs **535**
Lesson 6   Read Bar Graphs **541**

Wrap Up

My Review **547**
Reflect **550**
Performance Task **550PT1**

Look for this!
Click online and you can find tools that will help you explore concepts.

Chapter 8 Measurement and Time

Measurement and Data

ESSENTIAL QUESTION
How do I determine length and time?

Getting Started

Chapter 8 Project 552
Am I Ready? 553
My Math Words 554
My Vocabulary Cards 555
My Foldable **FOLDABLES** 561

Lessons and Homework

Lesson 1   Compare Lengths 563
Lesson 2   Compare and Order Lengths 569
Lesson 3   Nonstandard Units of Length 575
Lesson 4   Problem-Solving Strategy:
            Guess, Check, and Revise 581
Check My Progress 587
Lesson 5   Time to the Hour: Analog 589
Lesson 6   Time to the Hour: Digital 595
Lesson 7   Time to the Half Hour: Analog 601
Lesson 8   Time to the Half Hour: Digital 607
Lesson 9   Time to the Hour and Half Hour 613

Wrap Up

My Review 619
Reflect 622
Performance Task 622PT1

My classroom is fun!

connectED.mcgraw-hill.com

Chapter

9 Two-Dimensional Shapes and Equal Shares

Geometry

ESSENTIAL QUESTION
How can I recognize two-dimensional shapes and equal shares?

Getting Started

Chapter 9 Project **624**
Am I Ready? **625**
My Math Words **626**
My Vocabulary Cards **627**
My Foldable **FOLDABLES** **633**

Lessons and Homework

Lesson 1   Squares and Rectangles **635**
Lesson 2   Triangles and Trapezoids **641**
Lesson 3   Circles **647**
Lesson 4   Compare Shapes **653**
Check My Progress **659**
Lesson 5   Composite Shapes **661**
Lesson 6   More Composite Shapes **667**
Lesson 7   Problem-Solving Strategy:
            Use Logical Reasoning **673**
Check My Progress **679**
Lesson 8   Equal Parts **681**
Lesson 9   Halves **687**
Lesson 10   Quarters and Fourths **693**

Wrap Up

My Review **699**
Reflect **702**
Performance Task **702PT1**

Chapter

10 Three-Dimensional Shapes

Geometry

ESSENTIAL QUESTION
How can I identify three-dimensional shapes?

Getting Started

Chapter 10 Project **704**
Am I Ready? **705**
My Math Words **706**
My Vocabulary Cards **707**
My Foldable **FOLDABLES** **709**

Lessons and Homework

Lesson 1   Cubes and Prisms **711**
Lesson 2   Cones and Cylinders **717**
Check My Progress **723**
Lesson 3   Problem-Solving Strategy: Look for a Pattern **725**
Lesson 4   Combine Three-Dimensional Shapes **731**

Wrap Up

My Review **737**
Reflect **740**
Performance Task **740PT1**

Look for this!
Click online and you can get more help while doing your homework.
eHelp

connectED.mcgraw-hill.com

Chapter 1

Addition Concepts

ESSENTIAL QUESTION
How do you add numbers?

We're Going Outdoors!

Watch a video!

Watch

Chapter 1 Project

Treasure Hunt

1. Work with your group to solve the first problem. Check your answer at the location given to you by your teacher. Follow the clue to the next location.

2. Solve the second problem. Check your answer. Follow the clue to the next location.

3. Repeat the steps until your group has solved all four problems correctly.

I. Andrew has 4 cats. Zada has 0 cats. How many cats do they have all together?

_____ cats

2. Kylie has 3 hamsters. Nate has the same number of hamsters. How many hamsters do they have?

_____ hamsters

3. There are 10 cows and pigs on a farm. Write a way that could show the number of cows and pigs on the farm.

_____ + _____ = 10 cows and pigs

4. There are 5 fish in a tank. 4 snails are also in the tank. How many animals are in the tank?

_____ animals

Name _____

Am I Ready?

Write how many.

1.  _____

2.  _____

Draw circles to show each number.

3. **5**

4. **7**

Write how many there are in all.

5.

_____ birds

Shade the boxes to show the problems you answered correctly.

| 1 | 2 | 3 | 4 | 5 |

My Math Words

Vocab

Review Vocabulary

in all   same

Trace the words. Then draw a picture in each
box to show what each word means.

Word Set   **My Example**

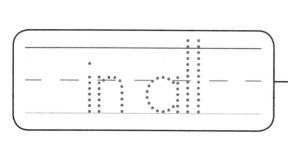

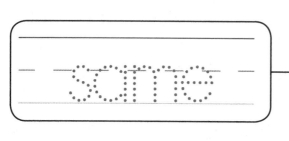

My Vocabulary Cards

 Vocab Processes &Practices

Lesson 1-2

add

0 1 2 3 4 5 6 ⑦ 8 9 10

7 + 3 = 10

Lesson 1-3

addition number sentence

4 + 4 = 8

9 = 6 + 3

Lesson 1-3

equals (=)

2 + 1 = 3

Lesson 1-13

false

3 + 1 = 5 is false

Lesson 1-2

part

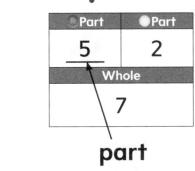

Part	Part
5	2
Whole	
7	

part

Lesson 1-3

plus (+)

4 + 2 = 6

Teacher Directions:
Ideas for Use

- Have students think of words that rhyme with some of the cards.
- Ask pairs of students to sort the cards by the number of syllables on each card.

- Have students draw examples for each card. Have them make drawings that are different from what is shown on the front of each card.

An expression using numbers and the + and = sign.

To join together sets to find the total or sum.

Something that is not a fact. The opposite of true.

The sign used to show having the same value as or is the same as.

The symbol used to show addition.

One of the two parts joined to make the whole.

My Vocabulary Cards

Vocab abc

Processes & Practices

✂

Lesson 1-3

sum

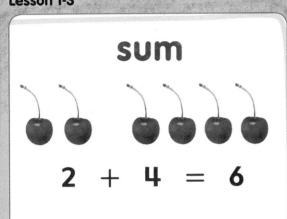

2 + 4 = 6

Lesson 1-13

true

4 + 1 = 5 is true

Lesson 1-2

whole

●Part	●Part
3	4
Whole	
7	

↑

whole

Lesson 1-4

zero

2 apples   0 apples

Teacher Directions:
More Ideas for Use

• Use the blank cards to write your own vocabulary words.

• Have students draw examples for each card. Have them make drawings that are different from what is shown on each card.

Something that is a fact. The opposite of false.

The answer to an addition problem.

A count of no objects.

The sum of two parts.

My Foldable

FOLDABLES® Follow the steps on the back to make your Foldable.

✂ -

Part	+	Part	=	Whole

$$1 + 3 = \boxed{}$$

$$2 + 6 = \boxed{}$$

$$6 + 1 = \boxed{}$$

$$3 + 6 = \boxed{}$$

$$\boxed{} + \boxed{} = 10$$

Study Organizer

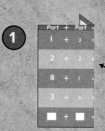

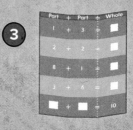

Name

Lesson 1
Addition Stories

ESSENTIAL QUESTION
How do you add numbers?

Math in My World

Teacher Directions: Use ⬤◯ to model. 3 children are swinging. 1 child is on the slide. How many children are at the park in all? Write the number.

Guided Practice

There are 4 ducks in the pond. 4 more ducks
walk to the pond. How many ducks are there in all?

_____8_____ ducks

Tell a number story. Use . Write how many in all.

1.

_____ turtles

2.

_____ birds

Talk Math   Tell how you put groups together.

Independent Practice

Tell a number story. Use 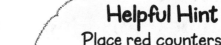.
Write how many in all.

Helpful Hint
Place red counters on the first group and place yellow counters on the second group.

3.

_____ foxes

4.

_____ deer

5.

_____ crabs

Problem Solving

Draw a picture to solve.

6. There are 6 gray cats. There are 3 black cats. How many cats are there in all?

_____ cats

Brain Builders

7. Ryan has 3 flashlights. Jade has the same number of flashlights as Ryan. How many flashlights do they have?

_____ flashlights

Write Math How did you find how many flashlights Ryan and Jade have?

_____

_____

_____

_____

_____

_____

Chapter I • Lesson I

Name

_____

My Homework

Homework Helper

Need help? connectED.mcgraw-hill.com

There are 3 marshmallows on one plate.
There are 2 marshmallows on the other plate.
How many marshmallows are there in all?

5 marshmallows

Practice

Tell a number story. Write how many there are in all.

1.

_____ sticks

2.

_____ s'mores

Draw a picture to solve.

3. Ella has 5 carrots. She gets 3 more. How many carrots does Ella have in all?

_____ carrots

Brain Builders

4. Joe has 4 trading cards. Kenji gives him the same amount. How many trading cards does Joe have now?

_____ trading cards

Explain how you solved the problem to a family member or friend.

5. **Test Practice** How many peppers are there in all?

5   7   8   9
○   ○   ○   ○

at Home Tell addition stories to your child. Have your child use buttons
odel the stories.

Name _____

Lesson 2
Model Addition

ESSENTIAL QUESTION ?
How do you add numbers?

 Math in My World Watch ▶ Tools

⬤ Part	⬤ Part
_____	_____
Whole	
_____	

 Teacher Directions: Use ⬤◯ to model. 2 girls bought tents from a store. I boy bought a tent from the same store. How many people bought tents in all? Write the numbers. Trace your counters to show the number of people who bought tents.

Online Content at ⚡ **connectED.mcgraw-hill.com**

Chapter I • Lesson 2   17

Copyright © McGraw-Hill Education.   ©Comstock Images/Alamy

Guided Practice

To find the **whole**, you **add** the **parts**.

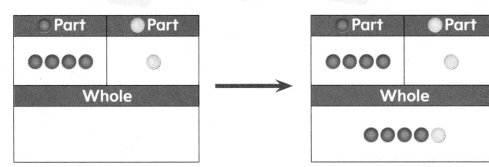

⬤ Part	⬤ Part
⬤⬤⬤⬤	⬤
Whole	

→

⬤ Part	⬤ Part
⬤⬤⬤⬤	⬤
Whole	
⬤⬤⬤⬤⬤	

⬤ Part	⬤ Part
4	1
Whole	

→

⬤ Part	⬤ Part
4	1
Whole	
5	

Use Work Mat 3 and ⬤⬤ to add.

1.

⬤ Part	⬤ Part
2	1
Whole	
___	

2.

⬤ Part	⬤ Part
5	3
Whole	
___	

3.

⬤ Part	⬤ Part
4	3
Whole	
___	

4.

⬤ Part	⬤ Part
2	4
Whole	
___	

Talk Math   How do you use ⬤⬤ to add 7 and 1?

Independent Practice

Use Work Mat 3 and ◐◯ to add.

5.

◐ Part	◯ Part
3	2
Whole	
_____	

6.

◐ Part	◯ Part
4	5
Whole	
_____	

7.

◐ Part	◯ Part
6	2
Whole	
_____	

8.

◐ Part	◯ Part
5	2
Whole	
_____	

9.

◐ Part	◯ Part
1	3
Whole	
_____	

10.

◐ Part	◯ Part
4	2
Whole	
_____	

11.

◐ Part	◯ Part
1	2
Whole	
_____	

12.

◐ Part	◯ Part
3	3
Whole	
_____	

Problem Solving

Use Work Mat 3 and ● ○ **if needed.**

13. Cristian saw 6 deer in a field.
Camila saw 2 deer in the forest.
How many deer did they see in all?

_____ deer

⚙ Brain Builders

14. Erin picks 6 flowers. Cliff picks some and
gives them to Erin. Now she has 7 flowers.
How many flowers did Cliff give Erin?

_____ flower(s)

Explain how you found the answer to a friend.

Write Math How do you find the whole?
Explain.

_____

_____

_____

_____

_____

_____

Name

My Homework

Homework Helper

Need help? connectED.mcgraw-hill.com

To find the whole, add the parts.

Part	Part
🪙🪙🪙	🪙🪙🪙🪙
Whole	
🪙🪙🪙🪙🪙🪙🪙	

Part	Part
3	4
Whole	
7	

Practice

Use pennies to add. Write the number.

1.

Part	Part
8	1
Whole	
_____	

2.

Part	Part
5	2
Whole	
_____	

3.

Part	Part
2	3
Whole	
_____	

4.

Part	Part
3	5
Whole	
_____	

Use pennies to add. Write the number.

5.

Part	Part
I	4
Whole	
_____	

6.

Part	Part
5	I
Whole	
_____	

Brain Builders

Draw a picture to solve the problem.

7. Stacie caught 4 fish in the morning. She caught one less fish in the afternoon than the morning. How many fish did Stacie catch?

_____ fish

Vocabulary Check

Complete the sentence.

in all   part

8. You can add the numbers from each _____ to find the whole.

Math at Home Place 2 red paper circles and 4 yellow paper circles on a table. Have your child count the circles. Ask your child to identify all of the different ways to make 6.

Name

Lesson 3
Addition Number Sentences

ESSENTIAL QUESTION ❓
How do you add numbers?

👐 Math in My World   ▶ Watch   🔧 Tools

_____ + _____ = _____
↳ Write your addition sentence here.

 Teacher Directions: Use 🧊 to model. There are 4 children playing tag. 2 more children join them. How many children are playing tag in all? Trace the cubes. Write the addition number sentence.

Guided Practice

You can write an addition number sentence.

See

Say   3   **plus**   2   **equals** ___5___

$\uparrow$

sum

Write ___3___   **+** ___2___   **=** ___5___

3 + 2 = 5 is an **addition number sentence**.

Write an addition number sentence.

1.

____ ◯ ____ ◯ ____

2.

____ ◯ ____ ◯ ____

3.

____ ◯ ____ ◯ ____

4.

____ ◯ ____ ◯ ____

Talk Math   What does the symbol + mean?

Copyright © McGraw-Hill Education   (l to r - t to b)(2)©Alex Cao/Photodisc/Getty Images; (3)©C Squared Studios/Photodisc/Getty Images; (4)©Ingram Publishing/SuperStock; (5, 6)©Jules Frazier/Photodisc/Getty Images; (7, 8)©Siede Preis/Photodisc/Getty Images; (9, 10)©Ken Karp/McGraw-Hill Education

Independent Practice

Write an addition number sentence.

5.

____ ◯ ____ ◯ ____

6.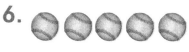

____ ◯ ____ ◯ ____

7.

____ ◯ ____ ◯ ____

8.

____ ◯ ____ ◯ ____

9.

____ ◯ ____ ◯ ____

10.

____ ◯ ____ ◯ ____

11.

____ ◯ ____ ◯ ____

12.

____ ◯ ____ ◯ ____

13. There are 2 dogs playing. 3 more dogs join them. How many dogs are playing in all?

_____ ◯ _____ ◯ _____ dogs

Brain Builders

14. Suzy sees 4 bees flying around a flower. Zoe sees one more bee than Suzy sees. How many bees did Suzy and Zoe see in all?

_____ bees

Write Math   What does = mean?

_____

- -

_____

- -

_____

- -

_____

Name _____

My Homework

Homework Helper 🏠 eHelp

Need help? connectED.mcgraw-hill.com

You can write an addition number sentence.

2 + 4 = 6

Practice

Write an addition number sentence.

1.

___ ◯ ___ ◯ ___

2.

___ ◯ ___ ◯ ___

3.

___ ◯ ___ ◯ ___

4.

___ ◯ ___ ◯ ___

Write an addition number sentence.

5.

6.

7. There are 5 cats at the park. 2 more join them. How many cats are there in all?

 cats

Brain Builders

8. There are 4 squirrels in a tall tree. There is one more squirrel than that in a short tree. How many squirrels are in both trees?

_____ squirrels

Vocabulary Check

Draw lines to match.

9. **addition number sentence**   The answer to an addition problem.

10. **sum** $4 + 5 = 9$

 Math at Home Create addition stories using cans of fruits or vegetables. Have your child write addition number sentences for the stories.

Name _____

ESSENTIAL QUESTION ❓
How do you add numbers?

 Math in My World

Lemonade for Sale

_____ + _____ = _____

↶ Write your addition sentence here.

 Teacher Directions: Use 🎲 to model. 7 people bought a glass of lemonade in the afternoon. 0 people bought a glass of lemonade in the evening. How many glasses of lemonade were sold in all? Write the addition number sentence.

Online Content at 🡕 connectED.mcgraw-hill.com   Chapter 1 • Lesson 4   29

Guided Practice

When you add **zero** to a number, the sum is the same as the number.

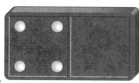

$4 + \mathbf{0} = \underline{4}$

sum

When you add a number to zero, the sum is the same as the number.

$\mathbf{0} + 2 = \underline{2}$

sum

Find each sum.

1.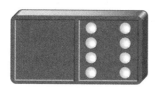

$0 + 8 = \underline{\hspace{1cm}}$

2.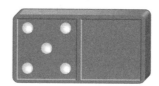

$5 + 0 = \underline{\hspace{1cm}}$

3.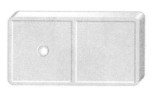

$1 + 0 = \underline{\hspace{1cm}}$

4.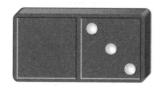

$0 + 3 = \underline{\hspace{1cm}}$

Talk Math What happens when you add zero to a number? Explain.

Independent Practice

Find each sum.

5.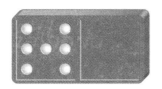

$7 + 0 = \underline{\hspace{2cm}}$

6.

$0 + 6 = \underline{\hspace{2cm}}$

7. $3 + 1 = \underline{\hspace{2cm}}$

8. $8 + 0 = \underline{\hspace{2cm}}$

9. $3 + 0 = \underline{\hspace{2cm}}$

10. $2 + 3 = \underline{\hspace{2cm}}$

11. $0 + 9 = \underline{\hspace{2cm}}$

12. $0 + 5 = \underline{\hspace{2cm}}$

13. $1 + 3 = \underline{\hspace{2cm}}$

14. $0 + 2 = \underline{\hspace{2cm}}$

15. $4 + 2 = \underline{\hspace{2cm}}$

Problem Solving

16. Jackson has 4 canoe paddles.
Ian has 0 paddles. How many
paddles do they have all together?

_____ paddles

Brain Builders

17. Grayson ate 3 hot dogs. Dre ate some
hot dogs. They ate 9 hot dogs in all.
How many hot dogs did Dre eat?

_____ hot dogs

Dre ate 0 more hot dogs that night. How many
hot dogs did Dre eat all day and night?

_____ hot dogs

18. Adrian adds 6 + 0 like this.
Tell why Adrian is wrong.
Make it right.

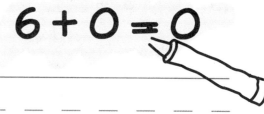

$$6 + 0 = 0$$

_____

\- \- \- \- \- \- \- \- \- \- \- \- \- \- \- \- \-

_____

\- \- \- \- \- \- \- \- \- \- \- \- \- \- \- \- \-

_____

\- \- \- \- \- \- \- \- \- \- \- \- \- \- \- \- \-

Name _____

My Homework

Homework Helper

Need help? connectED.mcgraw-hill.com

When you add zero, you add none.

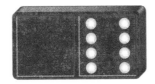   

$0 + 8 = 8$ $8 + 0 = 8$

Practice

Find each sum.

1.

   $9 + 0 =$ _____

2.

   $6 + 0 =$ _____

3.

   $0 + 4 =$ _____

4.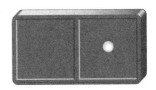

   $0 + 1 =$ _____

Find each sum.

5. 0 + 5 = _____

6. 3 + 5 = _____

7. 2 + 2 = _____

8. 0 + 9 = _____

9. There are 8 canoes on the lake.
   There are 0 canoes on the land.
   How many canoes are there in all?

_____ canoes

Brain Builders

10. There are 5 apples in a box. There
    are 0 apples in a bag. How many
    apples are there all together?

_____ apples

Tell a family member how you solved
the problem.

Vocabulary Check

Circle the correct number.

11. **zero**   1   6   0

 Math at Home Hold some cereal in one hand. Hold no cereal in another hand. Hold both hands out to your child. Ask your child to tell you which hand has zero pieces of cereal in it. Have your child add the cereal in each hand. Have your child say how many pieces of cereal there are in all.

Name _____

Check My Progress

Vocabulary Check

Draw lines to match.

1. =   addition number sentence

2. +   equals

3. 2 + 3 = 5   plus

4. 0   zero

Circle the correct answer.

5. When you _____ numbers together, you find the sum.

   zero   **add**

6. You can find the whole by adding the _____.

   parts   **sum**

7. Add two parts to find the _____.

   whole   **equals**

Concept Check

Add. Write the number.

8.

⬤Part	⬤Part
2	4
Whole	
_____	

9.

⬤Part	⬤Part
3	5
Whole	
_____	

Write an addition number sentence.

10.

_____ ◯ _____ ◯ _____

⚙ Brain Builders

11. There are 4 red apples and that same number of green apples in a bag. How many apples are there in the bag?

_____ ◯ _____ ◯ _____ apples

---

12. **Test Practice** Find the addition number sentence that matches.

$5 + 2 = 7$ $5 + 1 = 6$ $4 + 2 = 6$ $4 + 3 = 7$

◯   ◯   ◯   ◯

Name _____

Lesson 5
Vertical Addition

Math in My World   Watch ▶   Tools

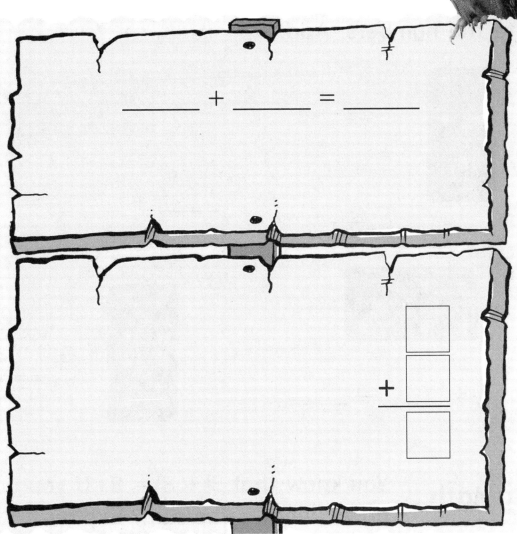

_____ + _____ = _____

+

Teacher Directions: Use ⬤◯ to model. Show 2 + 1 on each sign.
Trace the counters. Write the addition number sentences.

Guided Practice

You can add across. You can add down. The sum is the same if the numbers added are the same.

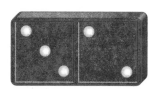

__3__ + __2__ = __5__

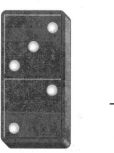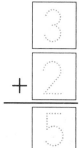

Write the numbers. Add.

1.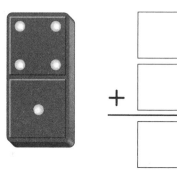

2.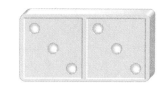

____ + ____ = ____

3.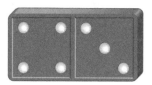

____ + ____ = ____

4.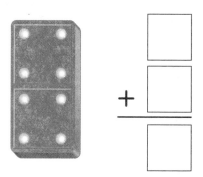

Talk Math You know that 5 + 3 = 8. If you add down, what is the sum? Explain.

Independent Practice

Write the numbers. Add.

5.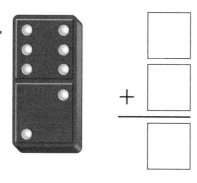

$$+$$

6.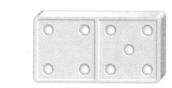

_____ + _____ = _____

7.

_____ + _____ = _____

8.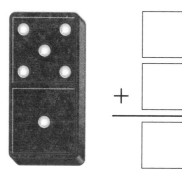

$$+$$

9.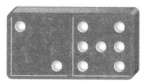

_____ + _____ = _____

10.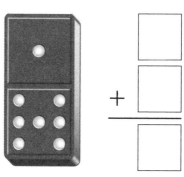

$$+$$

Add.

11. $2 + 6 = $ _____

12. $\begin{array}{r} 4 \\ + 5 \\ \hline \end{array}$

13. $1 + 3 = $ _____

Problem Solving

14. Brian saw 5 foxes in a field. He saw 3 other foxes in the woods. How many foxes did Brian see in all?

$$+ \boxed{}$$

_____ foxes

Brain Builders

15. Nate found 4 bugs. Pablo found I fewer bug than Nate. How many bugs did they find?

_____ bugs

Write Math How is adding down different from adding across? Explain.

_____

_ _

_____

_ _

_____

_ _

Name _____

My Homework

Homework Helper

Need help? connectED.mcgraw-hill.com

You can add across or you can add down.

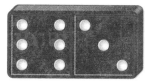

6 + 3 = 9

$$\begin{array}{r} 6 \\ + 3 \\ \hline 9 \end{array}$$

Practice

Write the numbers. Add.

1.

   _____ + _____ = _____

2.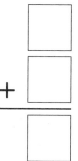

Add.

3. 2 + 6 = _____   4. 2 + 2 = _____   5. 1 + 4 = _____

Add.

6.　　2
　　　+ 7
　　　‾‾‾‾

7.　2 + 3 = ____

8.　　　I
　　　+ 6
　　　‾‾‾‾

9. There are 2 birds in a nest.
2 more birds fly to the nest.
How many birds are there
in all?

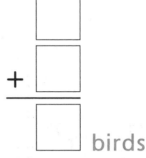

birds

Brain Builders

10. There are 5 children hiking. That same
number of children join the group. Now
how many children are hiking?

_____ children

11. **Test Practice** Which problem has the sum of 9?

○　　　　　○　　　　　○　　　　　○

　　4　　　　　7　　　　　4　　　　　2
　+ 5　　　　+ 3　　　　+ 4　　　　+ 6
　‾‾‾‾　　　　‾‾‾‾　　　　‾‾‾‾　　　　‾‾‾‾

Math at Home Give your child an addition number sentence. Have your child
show how to add across and down.

Name _____

Lesson 6
Problem Solving

STRATEGY: Write a Number Sentence

2 children are fishing.
4 more children join them.
How many children are fishing in all?

1 **Understand** Underline what you know.
Circle what you need to find.

2 **Plan** How will I solve the problem?

3 **Solve** I will write a number sentence.

$$2 + 4 = 6$$

_____ children are fishing in all.

4 **Check** Is my answer reasonable? Explain.

Cassie saw 2 elk in a field.
Marta saw 5 elk in the forest.
How many elk do they see
all together?

1 **Understand** Underline what you know.
Circle what you need to find.

2 **Plan** How will I solve the problem?

3 **Solve** I will...

_____ ◯ _____ ◯ _____

They see _____ elk all together.

4 **Check** Is my answer reasonable? Explain.

Apply the Strategy

Write an addition number sentence to solve.

I. Leon has 5 cards. Trey has 4 cards.
   How many cards do they have in all?

_____ ◯ _____ ◯ _____ cards

2. Nicki has 6 stickers. She was
   given 2 more stickers. How many
   stickers does she have now?

_____ ◯ _____ ◯ _____ stickers

Brain Builders

3. Isi saw 5 cars going down the road.
   Jamaal saw I fewer car. How many
   cars did Isi and Jamaal see?

_____ ◯ _____ ◯ _____ cars

Choose a strategy
- Write a number sentence.
- Make a table.
- Act it out.

4. Jayla and Will each catch 4 fish. How many total fish do Jayla and Will catch?

_____ fish

5. Deon has 4 jump ropes. Karen has 3 jump ropes. How many jump ropes do they have in all?

_____ jump ropes

6. There are 5 yellow beads and 4 red beads on a necklace. How many beads are on the necklace in all?

_____ beads

Name _____

My Homework

Homework Helper

Need help? connectED.mcgraw-hill.com

There are 2 birds singing.
4 more birds begin singing.
How many birds are singing in all?

1 Understand Underline what you know.
Circle what you need to find.

2 Plan How will I solve the problem?

3 Solve I will write a number sentence.

__2__ + __4__ = __6__ birds

4 Check Is my answer reasonable?

Problem Solving

Underline what you know. Circle what you need to find. Write an addition number sentence.

1. Mandy saw 3 rabbits. Tia saw 6 other rabbits. How many rabbits did they see in all?

_____ ◯ _____ ◯ _____ rabbits

2. There are 2 geese swimming in a pond. 4 more geese join them. How many geese are in the pond?

_____ ◯ _____ ◯ _____ geese

Brain Builders

3. Ben saw 5 raccoons. Conner saw 5 other raccoons. How many raccoons did they see in all? Explain how you know to a family member.

_____ ◯ _____ ◯ _____ raccoons

 Math at Home Take advantage of problem-solving opportunities during daily routines such as riding in the car, bedtime, doing laundry, putting away groceries, planning schedules, and so on.

Name _____

 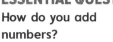
 Math in My World Watch ▶ Tools

_____ + _____ = _____

Write your addition sentence here.

 Teacher Directions: Use ● ○ to model. Mia's dad put 1 piece of wood on the fire. Then he put 3 more pieces on it. How many pieces of wood are on the fire in all? Trace the counters you used. Write the addition number sentence.

Online Content at connectED.mcgraw-hill.com

Guided Practice

There are different ways to make a sum of 4 and 5.

$2 + 2 =$ _____

$3 + 1 =$ _____

$2 + 3 =$ _____

$1 + 4 =$ _____

Use Work Mat 3 and ⚫⚪ to show different ways to make a sum of 4. Color the ◯. Write the numbers.

Ways to Make 4

1. ⚫◯◯◯   _____ + _____ = 4

2. ◯◯◯◯   _____ + _____ = 4

3. ◯◯◯◯   _____ + _____ = 4

Talk Math   What is another way to make 4?

Independent Practice

Use Work Mat 3 and ⬤◯ to show different ways to make a sum of 5. Color the ◯. Write the numbers.

Ways to Make 5

4. ◯◯◯◯◯ ____ + ____ = 5

5. ◯◯◯◯◯ ____ + ____ = 5

6. ◯◯◯◯◯ ____ + ____ = 5

7. ◯◯◯◯◯ ____ + ____ = 5

Add.

8. 1 + 4 = _____

9. 3 + 1 = _____

10. 3 + 2 = _____

11. 2 + 2 = _____

12.   1
    + 4

13.   4
    + 0

14.   0
    + 5

Problem Solving

Processes & Practices

Write an addition number sentence.

15. Kamie saw 3 turkeys. Taye saw 2 turkeys. How many turkeys did they see?

_____ + _____ = _____ turkeys

⚙ Brain Builders

16. Write a number sentence to show one way to make 4. Draw and color a picture that matches your number sentence.

_____ + _____ = _____

Write Math Is there more than one way to make 5? Explain.

_____

_ _ _ _ _ _ _ _ _ _ _ _ _ _ _ _ _ _

_____

_ _ _ _ _ _ _ _ _ _ _ _ _ _ _ _ _ _

_____

_ _ _ _ _ _ _ _ _ _ _ _ _ _ _ _ _ _

Name
....

My Homework

Homework Helper

Need help? connectED.mcgraw-hill.com

There are different ways to make a sum of 4 or 5.

⚫⚫⚪⚪ $2 + 2 = 4$

⚫⚪⚪⚪ $1 + 3 = 4$

⚫⚫⚫⚪⚪ $3 + 2 = 5$

⚫⚫⚫⚫⚪ $4 + 1 = 5$

Practice

Write different ways to make 4.

1. ⚫⚫⚫⚪   _____ + _____ = 4

2. ⚫⚫⚪⚪   _____ + _____ = 4

3. ⚫⚪⚪⚪   _____ + _____ = 4

4. ⚪⚪⚪⚪   _____ + _____ = 4

Write different ways to make 5.

5. ⚫⚫⚫⚫⚪ ____ + ____ = 5

6. ⚫⚫⚫⚪⚪ ____ + ____ = 5

7. ⚫⚫⚪⚪⚪ ____ + ____ = 5

8. ⚫⚪⚪⚪⚪ ____ + ____ = 5

9. ⚪⚪⚪⚪⚪ ____ + ____ = 5

10. Jose saw 3 green frogs and I red frog. How many frogs did he see in all?

_____ frogs

Brain Builders

11. **Test Practice** Jess drew 2 rainbows. Joe drew one more rainbow than Jess. How many rainbows did they draw?

2   3   5   6
○   ○   ○   ○

Math at Home Give your child five objects. Have your child show different ways to make 5.

Name _____

ESSENTIAL QUESTION ?
How do you add numbers?

 **Math in My World**

$\underline{\hspace{1cm}} + \underline{\hspace{1cm}} = \underline{\hspace{1cm}}$

Write your addition sentence here.

 Teacher Directions: Use ●○ to model. Find as many ways as you can to make 6 and 7. Trace counters above the tent to show one of those ways. Write the addition number sentence.

Guided Practice

There are many ways to make a sum of 6 and 7.

$1 + 5 = \underline{6}$

$2 + 4 = \underline{6}$

$3 + 4 = \underline{7}$

$5 + 2 = \underline{7}$

Use Work Mat 3 and ⬤◯ to show different ways to make a sum of 6. Write the numbers.

Ways to Make 6

1. _____ + _____ = 6

2. _____ + _____ = 6

3. _____ + _____ = 6

4. _____ + _____ = 6

Helpful Hint
Think of all of the different ways to make 6.

Talk Math   Is 5 + 1 the same as 4 + 2? Explain.

Independent Practice

Use Work Mat 3 and ⬤◯ to show different ways to make a sum of 7. Write the numbers.

Ways to Make 7

5. _____ + _____ = 7   6. _____ + _____ = 7

7. _____ + _____ = 7   8. _____ + _____ = 7

9. _____ + _____ = 7   10. _____ + _____ = 7

Add.

11. $4 + 2 =$ _____   12. $3 + 4 =$ _____

13. $7 + 0 =$ _____   14. $3 + 3 =$ _____

15. $4 + 3 =$ _____   16. $5 + 1 =$ _____

17.   1   18.   0   19.   5
   + 6   + 6   + 2

Problem Solving

Processes & Practices

Write an addition number sentence.

20. Victoria caught 5 fish. Her brother caught 2 fish. How many fish did they catch all together?

_____ + _____ = _____ fish

Brain Builders

21. There are 4 ducks swimming in a pond. 3 more ducks join them. How many ducks are swimming in the pond? Explain your answer to a friend.

_____ + _____ = _____ ducks

22. Mason added 3 + 3 like this. Tell why Mason is wrong. Make it right.

$$3 + 3 = 7$$

_____

_ _ _ _ _ _ _ _ _ _ _ _ _ _ _ _ _ _

_____

_ _ _ _ _ _ _ _ _ _ _ _ _ _ _ _ _ _

_____

_ _ _ _ _ _ _ _ _ _ _ _ _ _ _ _ _ _

_____

Name ....

My Homework

Homework Helper

Need help? connectED.mcgraw-hill.com

There are many ways to make sums of 6 and 7.

$4 + 2 = 6$ $5 + 2 = 7$

$3 + 3 = 6$ $3 + 4 = 7$

Practice

Write different ways to make 6 and 7.

Ways to Make 6 or 7

1. _____ + _____ = 6   2. _____ + _____ = 6

3. _____ + _____ = 6   4. _____ + _____ = 6

5. _____ + _____ = 7   6. _____ + _____ = 7

7. _____ + _____ = 7   8. _____ + _____ = 7

Add.

9. 6
    + 0

10. 5
    + 2

11. 4
    + 3

12. 1
    + 6

13. 3
    + 3

14. 2
    + 4

15. There are 2 rabbits eating. 5 more rabbits join them. How many rabbits are eating in all?

_____ rabbits

 Brain Builders

16. **Test Practice** Lily and Kyle each planted 3 trees. How many trees were planted?

8 trees   7 trees   6 trees   4 trees

○   ○   ○   ○

 Math at Home Give your child 7 objects. Then have your child show different ways to make two groups that show 7 in all.

Name

_____

Lesson 9
Ways to Make 8

ESSENTIAL QUESTION
How do you add numbers?

 Math in My World Watch Tools

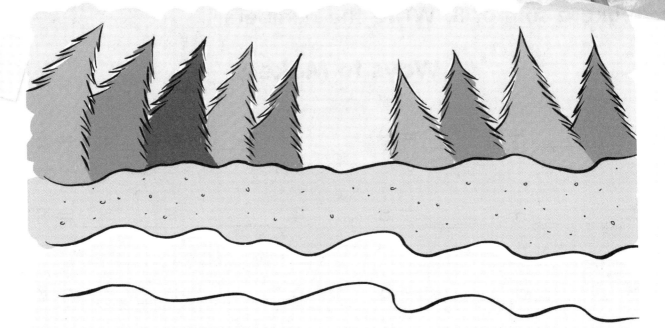

$$\underline{} + \underline{} = \underline{}$$
Write your addition sentence here.

 Teacher Directions: Use to model. There are 5 people swimming. 3 more people join them. How many people are swimming in all? Write the addition number sentence.

Guided Practice

There are many ways to make sums of 8.

3 + 5 = ___

1 + 7 = ___

Use Work Mat 3 and to show different ways to make a sum of 8. Write the numbers.

Ways to Make 8

1. _____ + _____ = 8   2. _____ + _____ = 8

3. _____ + _____ = 8   4. _____ + _____ = 8

5. _____ + _____ = 8   6. _____ + _____ = 8

Add.

7. 4 + 4 = _____   8. 6 + 2 = _____

9. 8 + 0 = _____   10. 7 + 1 = _____

Talk Math   How could you use counters to show ways to make 8?

Independent Practice

Use Work Mat 3 and ⚫⚪. Add.

11. $6 + 2 =$ _____   12. $0 + 3 =$ _____

13. $2 + 5 =$ _____   14. $2 + 6 =$ _____

15. $4 + 4 =$ _____   16. $0 + 7 =$ _____

17. $1 + 7 =$ _____   18. $3 + 5 =$ _____

19. $2 + 6 =$ _____   20. $8 + 0 =$ _____

21. $4 + 2 =$ _____   22. $7 + 1 =$ _____

23. $\begin{array}{r} 4 \\ + 3 \\ \hline \end{array}$ 24. $\begin{array}{r} 4 \\ + 4 \\ \hline \end{array}$ 25. $\begin{array}{r} 3 \\ + 1 \\ \hline \end{array}$

26. $\begin{array}{r} 2 \\ + 3 \\ \hline \end{array}$ 27. $\begin{array}{r} 6 \\ + 1 \\ \hline \end{array}$ 28. $\begin{array}{r} 5 \\ + 3 \\ \hline \end{array}$

Problem Solving

Write an addition number sentence.

29. Brice saw 5 foxes. His friend saw
3 other foxes. How many foxes
did they see in all?

_____ + _____ = _____ foxes

Brain Builders

30. There are 8 chipmunks. Some are on
the tree and some are on the ground.
How many chipmunks could be in the
tree and on the ground? Find as many
ways as you can.

_____ + _____ = __**8**__ chipmunks

31. A baker sold 4 muffins in the morning. She
sold 2 muffins later that day. The answer
is 6 muffins. What is the question?

_____

_____

_____

_____

Name ....

My Homework

Homework Helper eHelp

Need help? connectED.mcgraw-hill.com

There are many ways to make a sum of 8.

⬤⬤⬤⬤⬤◯◯◯   5 + 3 = 8

⬤⬤◯◯◯◯◯◯   2 + 6 = 8

Practice

Write different ways to make 8.

1. _____ + _____ = 8   2. _____ + _____ = 8

3. _____ + _____ = 8   4. _____ + _____ = 8

5. _____ + _____ = 8   6. _____ + _____ = 8

Add.

7. 3 + 5 = _____   8. 8 + 0 = _____

Add.

9. 6
    + 2
    ———

10. 3
    + 5
    ———

11. 4
    + 3
    ———

12. 4
    + 4
    ———

13. 6
    + 1
    ———

14. 1
    + 7
    ———

15. 5 people went canoeing. 2 more people joined them. How many people went canoeing in all?

_____ + _____ = _____ people

16. **Test Practice** Which number sentence has a sum of 8?

6 + 1   3 + 4   0 + 5   3 + 5
  ○   ○   ○   ○

Math at Home Give your child 8 objects. Then have your child show different ways to make a sum of 8.

Name

Vocabulary Check

Circle the correct answer.

add   **addition number sentence**   **sum**

1. A _____ is an answer to an addition problem.

Concept Check

Add. Write the numbers.

2.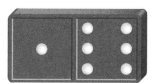

_____ + _____ = _____

3.

4.

5.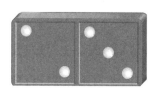

_____ + _____ = _____

Copyright © McGraw-Hill Education

Add.

6. 2 + 3 = _____   7. 2 + 2 = _____

8.   5   9.   6   10.   1
   + 3   + 2   + 5
   ———   ———   ———

11.   3   12.   1   13.   1
   + 4   + 7   + 3
   ———   ———   ———

Write an addition number sentence to solve.

14. There are 6 birds flying together. There
    is 1 other bird sitting on a tree branch.
    How many birds are there in all?

    _____ + _____ = _____ birds

Brain Builders

15. **Test Practice** There are 2 beavers swimming
    in a river. The same number of beavers join
    them. How many beavers are now swimming
    together?

    2 beavers   4 beavers   5 beavers   6 beavers
      ○   ○   ○   ○

Name _____

Lesson 10
Ways to Make 9

 Math in My World   Watch ▶ Tools 🎲

↱ _____ + _____ = _____
Write your addition sentence here.

 Teacher Directions: Use two crayons to color the stars to show a way to make 9. Write the addition number sentence.

Guided Practice

There are different ways to make a sum of 9.

 $1 + 8 =$ ___9___

 $2 + 7 =$ ___9___

$3 + 6 =$ ___9___

Use Work Mat 3 and ⬤◯ to show different ways to make a sum of 9. Write the numbers.

Ways to Make 9

1. ____ + ____ = 9   2. ____ + ____ = 9

3. ____ + ____ = 9   4. ____ + ____ = 9

5. ____ + ____ = 9   6. ____ + ____ = 9

7. ____ + ____ = 9   8. ____ + ____ = 9

Talk Math   Why do you get the same sum when you add $6 + 3$ and $7 + 2$?

Independent Practice

Use Work Mat 3 and **. Add.**

9. 5 + 4 = _____   10. 2 + 5 = _____

11. 4 + 2 = _____   12. 5 + 3 = _____

13. 2 + 7 = _____   14. 4 + 3 = _____

15. 3 + 6 = _____   16. 4 + 4 = _____

17. 8 + 1 = _____   18. 6 + 1 = _____

19. 1 + 4 = _____   20. 2 + 6 = _____

21.	22.	23.
7 + 1	0 + 9	1 + 8

24.	25.	26.
4 + 2	9 + 0	6 + 3

Problem Solving

Write an addition number sentence.

27. There are 3 yellow fish and 6 red fish in a pond. How many fish are there in all?

_____ + _____ = _____ fish

Brain Builders

28. There are 6 turtles. 4 are walking to the pond, and the others are sitting on a log. How many turtles are on the log?

_____ turtles

Write Math Is there more than one way to make 9? Explain.

_____

_____

_____

_____

Name _____

My Homework

Homework Helper   Need help? 🖰 connectED.mcgraw-hill.com

There are many ways to make a sum of 9.

  $3 + 6 = 9$

  $2 + 7 = 9$

Practice

Write different ways to make 9.

Ways to Make 9

1. _____ + _____ = 9

2. _____ + _____ = 9

3. _____ + _____ = 9

4. _____ + _____ = 9

5. _____ + _____ = 9

6. _____ + _____ = 9

Add.

7. $4 + 5 =$ _____

8. $2 + 7 =$ _____

Add.

9.   6
     + 3
     ─────

10.   5
     + 1
     ─────

11.   9
     + 0
     ─────

12.   4
     + 3
     ─────

13.   7
     + 2
     ─────

14.   5
     + 3
     ─────

15. There are 5 owls sitting on a tree.
    There are 4 other owls flying. How
    many owls are there in all?

    _____ owls

 Brain Builders

16. **Test Practice** Which is not a way to
    make a sum of 9?

    5 + 4   1 + 8   4 + 4   6 + 3
      ○   ○   ○   ○

 Math at Home Give your child 9 objects. Have your child show different ways
to make two groups to show 9.

Name ....

Ways to Make 10

ESSENTIAL QUESTION
How do you add numbers?

 Math in My World Watch Tools

$$\underline{\hspace{2cm}} + \underline{\hspace{2cm}} = 10$$

 Teacher Directions: Use ⚫⚪ to model. Find different ways to make 10. Trace and color counters to show one of the ways. Write the numbers.

Guided Practice

There are many ways to make 10.

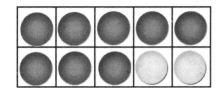

___8___ + ___2___ = 10

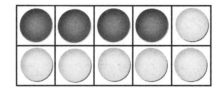

___4___ + ___6___ = 10

Write the numbers that make 10.

1.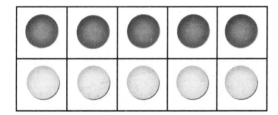

_____ + _____ = 10

2.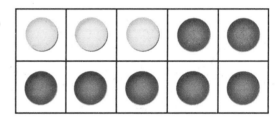

_____ + _____ = 10

3.

_____ + _____ = 10

4.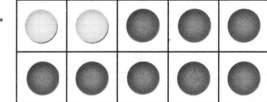

_____ + _____ = 10

Talk Math Name all the ways to make 10 on a ten-frame using 2 numbers.

Name _____

Write the numbers that make 10.

5.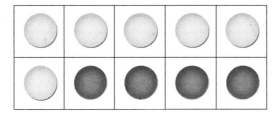

_____ + _____ = 10

6.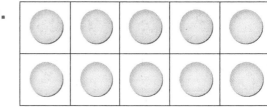

_____ + _____ = 10

7.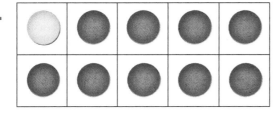

_____ + _____ = 10

8.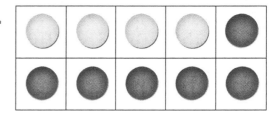

_____ + _____ = 10

Draw and color a way to make 10 using two numbers. Write the numbers.

9.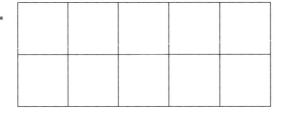

_____ + _____ = 10

10.

_____ + _____ = 10

Problem Solving

11. There are 5 red counters. How many yellow counters will make 10 in this ten-frame? Draw and color the counters.

_____ yellow counters

 Brain Builders

12. Joe saw 2 geese. How many more geese does he need to see so that he sees 10 geese in all?

_____ more geese

13. Riley wrote this on the board. Tell why Riley is wrong. Make it right.

$$6 + 5 = 10$$

_____

\- __ __ __ __ __ __ __ __ __ __ __ __ __ __ __ __

_____

\- __ __ __ __ __ __ __ __ __ __ __ __ __ __ __ __

_____

\- __ __ __ __ __ __ __ __ __ __ __ __ __ __ __ __

Name _____

My Homework

Homework Helper

Need help? ⟋ connectED.mcgraw-hill.com

There are many ways to make 10.

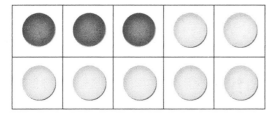

$$3 + 7 = 10 \qquad 6 + 4 = 10$$

Practice

Write the numbers that make 10.

1.

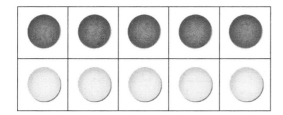

_____ + _____ = 10

2.

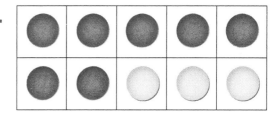

_____ + _____ = 10

3.

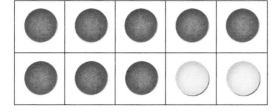

_____ + _____ = 10

4.

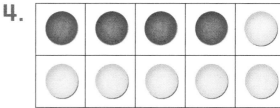

_____ + _____ = 10

Draw and color a way to make 10 using two numbers. Write the numbers.

5.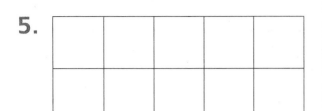

_____ + _____ = 10

6.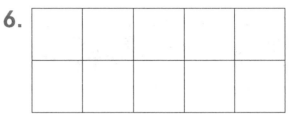

_____ + _____ = 10

Brain Builders

7. There are 3 bears drinking from a creek. How many more bears need to join them to make 10 bears drinking from the creek?

_____ bears

8. **Test Practice** Liam has 6 counters. How many more counters does he need to make 10 counters?

4 counters   5 counters   6 counters   16 counters
○   ○   ○   ○

 Math at Home Give your child ten crayons. Have him or her put the crayons in two groups showing a way to make 10. Ask your child to show another way to make 10.

Lesson 12
Find Missing Parts of 10

ESSENTIAL QUESTION
How do you add numbers?

 Math in My World   Watch Tools

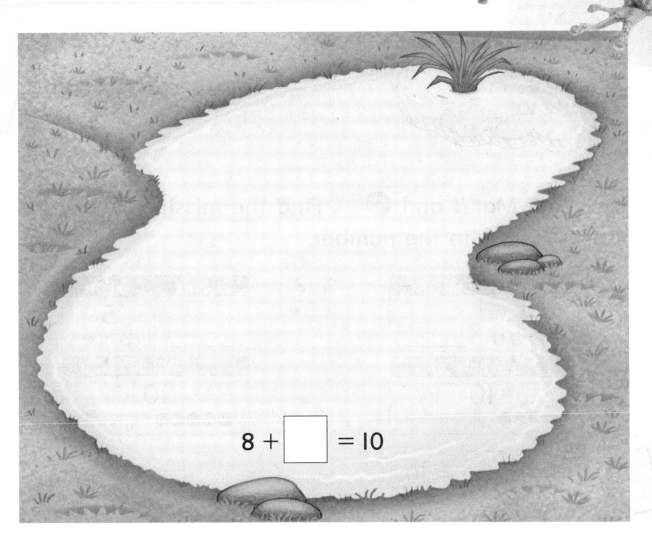

$$8 + \boxed{} = 10$$

 Teacher Directions: Use to model. There are 10 frogs in all. 8 of the frogs are in the pond. The rest are in the grass. How many frogs are in the grass? Write the missing part.

Online Content at connectED.mcgraw-hill.com

Guided Practice

The whole is 10. One part is 3.
What is the other part?

○ Part	● Part
●●●	_____
Whole	
●●●○○○○○○○	

○ Part	● Part
3	_____
Whole	
10	

Helpful Hint
You can use counters to find the missing part of 10.

$3 + \boxed{7} = 10$

Use Work Mat 3 and ●○. Find the missing
part of 10. Write the number.

1.

○ Part	● Part
4	_____
●●●●	
Whole	
10	
●●●●○○○○○○	

$4 + \boxed{} = 10$

2.

● Part	● Part
_____	5
	○○○○○
Whole	
10	
●●●●●○○○○○	

$\boxed{} + 5 = 10$

Talk Math If you know one of the parts and the
whole, how do you find the other part?

Independent Practice

Use Work Mat 3 and ⚫⚪. Find the missing part of 10. Write the number.

3.

⚫ Part	⚪ Part
7	_____
Whole	
10	

$$7 + \boxed{} = 10$$

4.

⚫ Part	⚪ Part
_____	5
Whole	
10	

$$\boxed{} + 5 = 10$$

5.

⚫ Part	⚪ Part
_____	8
Whole	
10	

$$\boxed{} + 8 = 10$$

6.

⚫ Part	⚪ Part
1	_____
Whole	
10	

$$1 + \boxed{} = 10$$

7.

⚫ Part	⚪ Part
3	_____
Whole	
10	

$$3 + \boxed{} = 10$$

8.

⚫ Part	⚪ Part
_____	6
Whole	
10	

$$\boxed{} + 6 = 10$$

Problem Solving

9. There are 10 apples. 3 of the apples are green. The rest of the apples are red. How many apples are red?

_____ apples

Brain Builders

10. Javier has 10 leaves. The leaves are orange and yellow. Write how many leaves could be orange and how many could be yellow.

_____ yellow   _____ orange

11. Angelo wrote the missing part. Tell why Angelo is wrong. Make it right.

Part	Part
2	7
Whole	
10	

Name

My Homework

Homework Helper

Need help? connectED.mcgraw-hill.com

You can find the missing part of 10.

⚫ Part	⚪ Part
6	_____
Whole	
10	

6 + 4 = 10

⚫ Part	⚪ Part
8	_____
Whole	
10	

8 + 2 = 10

Practice

Find the missing part of 10. Write the number.

1.

⚫ Part	⚪ Part
3	_____
Whole	
10	

3 + ☐ = 10

2.

⚫ Part	⚪ Part
_____	5
Whole	
10	

☐ + 5 = 10

Find the missing part of 10. Write the number.

3.

● Part	● Part
6	_____

Whole
10

$$6 + \boxed{} = 10$$

4.

● Part	● Part
_____	1

Whole
10

$$\boxed{} + 1 = 10$$

Brain Builders

5. Amaya sees 10 bugs on a log. The bugs are red and black. Write how many bugs could be red and how many could be black.

_____ red  _____ black

6. **Test Practice** Carter sees 10 eagles. 5 of the eagles are flying. The rest of the eagles are in their nest. How many eagles are in their nest?

15 eagles   9 eagles   10 eagles   5 eagles
  ○   ○   ○   ○

 Math at Home Create number cards with the numbers 0 to 10 on them. Show one number card. Have your child find the other number card that shows the number needed to make 10.

Name _____

 Math in My World Tools

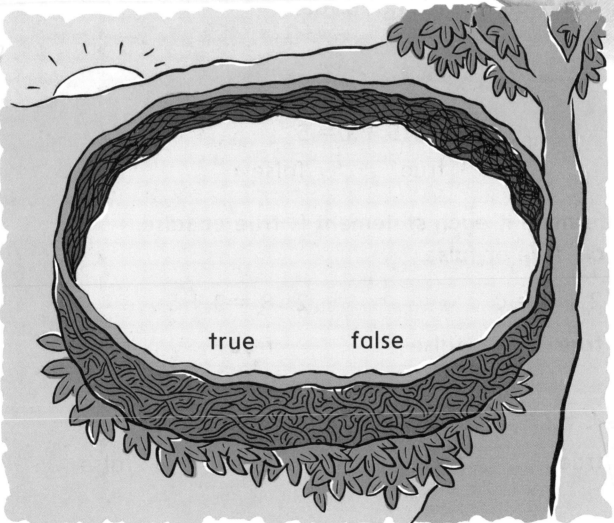

true   false

 Teacher Directions: Use ⬤ ◯ to model. There are 3 yellow and 3 red eggs in a nest. There are 7 eggs in all. Is this true or false? Circle the word. Trace the counters you used to show the problem.

Guided Practice

Statements can be true or false.

A **true** statement is a fact.

5 + 1 = 6

(true)   false

A **false** statement is not a fact.

5 + 1 = 5

true   (false)

**Determine if each statement is true or false.
Circle true or false.**

1. 2 + 4 = 6

   true   false

2. 8 = 3 + 5

   true   false

3. 1 + 7 = 9

   true   false

4. 7 = 7

   true   false

Talk Math   Tell your own false addition statement to a classmate.

Independent Practice

**Determine if each statement is true or false.
Circle true or false.**

5. 1 + 3 = 5

    true   false

6. 5 + 5 = 10

    true   false

7. 3 + 5 = 7

    true   false

8. 9 = 9 + 0

    true   false

9. 6 + 2 = 8

    true   false

10. 5 + 2 = 4

    true   false

11. 3 = 3

    true   false

12. 4 + 2 = 7

    true   false

13. 9 = 8 + 2

    true   false

14. 2 + 5 = 7

    true   false

15.
$$\begin{array}{r} 4 \\ +\ 4 \\ \hline 8 \end{array}$$

    true   false

16.
$$\begin{array}{r} 6 \\ +\ 1 \\ \hline 5 \end{array}$$

    true   false

Problem Solving

Determine if the word problem is true or false.
Circle true or false.

17. There are 4 children bird watching.
3 more children join them. There are
7 children bird watching in all.

true   false

Brain Builders

18. Write a true number sentence and a false
number sentence. Ask a friend to circle which
number sentence is true and which is false.

_____ = _____ + _____   _____ = _____ + _____
true   false   true   false

Write Math   Is 6 + 2 = 3 + 6 a true or false
math statement? Explain.

_____

– –

_____

– –

_____

– –

Name _____

My Homework

Homework Helper Need help? connectED.mcgraw-hill.com

Math statements are true or false.

$5 + 1 = 4$ $3 + 2 = 5$

true   (false)   (true)   false

Practice

Determine if each statement is true or false.
Circle true or false.

1. $3 + 1 = 4$

   true   false

2. $0 = 4 + 0$

   true   false

3. $5 + 4 = 9$

   true   false

4. $10 = 6 + 4$

   true   false

5. $3 + 6 = 10$

   true   false

6. $4 + 1 = 5$

   true   false

Determine if each statement is true or false. Circle true or false.

7. $9 = 8 + 1$

    true   false

8. $6 + 2 = 3 + 6$

    true   false

9. $$\begin{array}{r} 1 \\ + \ 0 \\ \hline 0 \end{array}$$

    true   false

10. $$\begin{array}{r} 3 \\ + \ 3 \\ \hline 6 \end{array}$$

    true   false

Brain Builders

11. Write a true or false number sentence. Circle if it is true or false.

_____ = _____ + _____

    true   false

Vocabulary Check

Draw lines to match.

12. **true**   Something that is not a fact.

13. **false**   Something that is a fact.

 Math at Home Tell your child a false addition number sentence. Ask your child if it is true or false. Have your child make it true.

Name _____

Processes
&Practices

Fluency Practice

Add.

1. 4 + 6 = _____

2. 5 + 4 = _____

3. 3 + 2 = _____

4. 2 + 6 = _____

5. 2 + 5 = _____

6. 1 + 3 = _____

7. 7 + 1 = _____

8. 0 + 9 = _____

9. 1 + 1 = _____

10. 3 + 7 = _____

11. 4 + 4 = _____

12. 5 + 1 = _____

13. 7 + 3 = _____

14. 2 + 7 = _____

15. 4 + 3 = _____

16. 3 + 0 = _____

17. 0 + 5 = _____

18. 8 + 2 = _____

19. 5 + 3 = _____

20. 9 + 1 = _____

21. 4 + 5 = _____

22. 1 + 2 = _____

23. 7 + 0 = _____

24. 3 + 5 = _____

Copyright © McGraw-Hill Education

Online Content at connectED.mcgraw-hill.com

Fluency Practice

Add.

1. $\begin{array}{r} 7 \\ + 2 \\ \hline \end{array}$ 2. $\begin{array}{r} 4 \\ + 3 \\ \hline \end{array}$ 3. $\begin{array}{r} 6 \\ + 2 \\ \hline \end{array}$ 4. $\begin{array}{r} 1 \\ + 8 \\ \hline \end{array}$

5. $\begin{array}{r} 2 \\ + 4 \\ \hline \end{array}$ 6. $\begin{array}{r} 4 \\ + 5 \\ \hline \end{array}$ 7. $\begin{array}{r} 2 \\ + 2 \\ \hline \end{array}$ 8. $\begin{array}{r} 5 \\ + 1 \\ \hline \end{array}$

9. $\begin{array}{r} 4 \\ + 6 \\ \hline \end{array}$ 10. $\begin{array}{r} 2 \\ + 7 \\ \hline \end{array}$ 11. $\begin{array}{r} 3 \\ + 3 \\ \hline \end{array}$ 12. $\begin{array}{r} 5 \\ + 3 \\ \hline \end{array}$

13. $\begin{array}{r} 6 \\ + 1 \\ \hline \end{array}$ 14. $\begin{array}{r} 2 \\ + 8 \\ \hline \end{array}$ 15. $\begin{array}{r} 0 \\ + 6 \\ \hline \end{array}$ 16. $\begin{array}{r} 1 \\ + 9 \\ \hline \end{array}$

17. $\begin{array}{r} 0 \\ + 4 \\ \hline \end{array}$ 18. $\begin{array}{r} 7 \\ + 3 \\ \hline \end{array}$ 19. $\begin{array}{r} 3 \\ + 5 \\ \hline \end{array}$ 20. $\begin{array}{r} 10 \\ + 0 \\ \hline \end{array}$

Name _____

My Review

Vocabulary Check

Complete each sentence.

| add | part | sum | whole |

1. Joining two parts together makes a _____.

2. A _____ is one of the groups that are joined when adding.

3. When you _____ two numbers together, you come up with the sum.

4. The answer in an addition number sentence is called the _____.

Concept Check

Write the addition number sentence.

5.

6.

Add.

7. 1 + 6 = ____

8. 8 + 0 = ____

9. 3 + 2 = ____

10. 4 + 4 = ____

11.   4
   + 0
   ——

12.   3
   + 1
   ——

13.   5
   + 4
   ——

14.   2
   + 6
   ——

Show a way to make the sum. Color the ◯.
Write the numbers.

15. ◯ ◯ ◯ ◯ ◯ ◯   ____ + ____ = 6

16. ◯ ◯ ◯ ◯ ◯   ____ + ____ = 5

Find the missing part of ten. Write the number.

17.

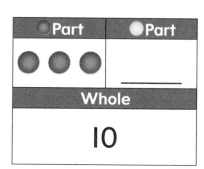

◯ Part	◯ Part
● ● ●	————
Whole	
10	

3 + [ ] = 10

18.

◯ Part	◯ Part
————	◯
Whole	
10	

[ ] + 1 = 10

Name ....

 ## Problem Solving

Write an addition number sentence.

19. Liliana finds 3 twigs. Enrique finds 2 twigs. How many twigs do they find in all?

_____ + _____ = _____ twigs

Determine if the statement is true or false. Circle true or false.

20. There are 3 bears yawning. 4 more bears begin yawning. There are 8 bears yawning in all.

true   false

Brain Builders

21. Test Practice Miranda sees 5 stars in the sky. Madison sees the same number of stars in the sky. How many stars do they see in all?

4 stars   5 stars   9 stars   10 stars
   ◯   ◯   ◯   ◯

Reflect

Show the ways to answer.

Find the whole.

⬤ Part	⬤ Part
3	5
Whole	
_____	

Write an addition number sentence.

_____ + _____ = _____

ESSENTIAL QUESTION

How do you add numbers?

Add.

```
    3
  + 4
  ___
  [ ]
```

Add zero.

$0 + 9 =$ ___

Now I Know!

Performance Task

Brain Builders

A Lemonade Stand

Marta is selling lemonade. She needs to buy lemons.

Show all your work to receive full credit.

Part A

Marta fills two bags with lemons. Write an addition sentence to show the number of lemons she buys.

_____ + _____ = _____ lemons

Part B

Marta sold 4 cups of lemonade in the morning and 3 cups in the afternoon. How many cups did Marta sell that day?

_____ + _____ = _____ cups

Part C

10 people smiled at Marta. 4 were boys. Find the missing part to find out how many were girls.

Part	Part
4 boys	_____ girls
Whole	
10 people	

Part D

Marta also gave away water for free. In the morning she gave away 0 cups of water. In the afternoon she gave away 6 cups of water. How many cups did she give away in all?

_____ + _____ = _____ cups of water

ESSENTIAL QUESTION

How do you subtract numbers?

Let's Go on a Safari!

Watch a video!

Watch

Chapter 2 Project

Subtraction Storybook

1. Decide how to design the cover of your subtraction storybook.

2. Write a title for your book.

3. Use drawings and words to design your cover for the book below.

Name _____

Write how many.

1.

2.

_____   _____

Draw circles to show each number.

3.   **6**

4.   **4**

Put an X on 2 frogs. Write how many are left.

5.   _____ frogs

Shade the boxes to show the problems you answered correctly.

How Did I Do?

1	2	3	4	5

Online Content at connectED.mcgraw-hill.com

My Math Words

Vocab
abc

Review Vocabulary

equals	join	take away

Trace the words. Then draw a picture in the box to show what the word means.

Word	**My Example**

equals

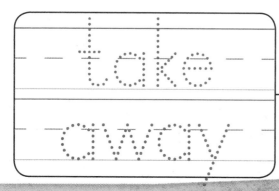

join

take
away

My Vocabulary Cards

Vocab

Processes & Practices

Lesson 2-7

compare

2 more

2 fewer

Lesson 2-3

difference

$4 - 1 = 3$

Lesson 2-3

minus (−)

$6 - 2 = 4$

Lesson 2-13

related facts

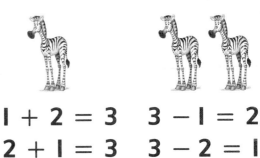

$1 + 2 = 3$ $3 - 1 = 2$
$2 + 1 = 3$ $3 - 2 = 1$

Lesson 2-2

subtract

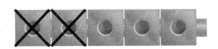

$5 - 2 = 3$

Lesson 2-3

subtraction number sentence

$5 - 2 = 3$

Teacher Directions:
Ideas for Use
• Have students draw different examples
  for each card.

• Have students arrange cards to show
  words with similar meanings. Have
  them explain the meaning of their
  grouping.

The answer to a subtraction problem.

Look at groups of objects, shapes, or numbers and see how they are alike or different.

Basic facts using the same numbers.

The sign used to show subtraction.

An expression using numbers and the − and = signs.

To take away.

My Vocabulary Cards

 Vocab

Teacher Directions:
More Ideas for Use

- Use the blank cards to write your own vocabulary words.

- Have students draw pictures on the blank cards to show the meaning of each new vocabulary word.

4 − 0

4 − 1

4 − 2

4 − 3

4 − 4

_____

_____

_____

_____

_____

Name

Lesson 1
Subtraction Stories

ESSENTIAL QUESTION
How do you subtract numbers?

 Math in My World Watch Tools

_____ dragonflies

 Teacher Directions: Use ⬤◯ to model. There are 7 dragonflies sitting on a flower. 2 fly away. How many dragonflies are left on the flower? Write the number.

Online Content at connectED.mcgraw-hill.com

Guided Practice

There are 6 cats on a fence.
I cat jumps down off of the fence.

How many cats are left on the fence? ___5___ cats

Tell a number story. Use ⬤◯. Write how many are left.

1.

How many birds are left in the bird bath? _____ birds

2.

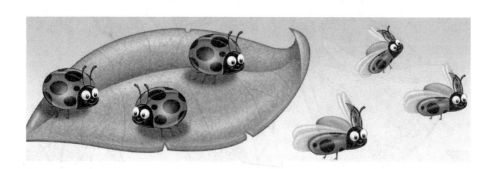

How many ladybugs are left on the leaf? _____ ladybugs

Talk Math How are addition and subtraction stories different?

Name ....

Independent Practice

Tell a number story. Use . Write how many are left.

3.

How many ants are left on the leaf? _____ ants

4.

How many butterflies are left on the bush?

_____ butterflies

5.

How many kites are left on the ground? _____ kites

Problem Solving

Use to solve.

ocesses
&Practices

6. There are 4 people hiking on a trail. 2 of the people go home. How many people are left?

_____ people

 Brain Builders

7. There are 8 bees near the hive. 3 bees fly away. Then another bee flies away. How many bees are left?

_____ bees

8. There are 6 tigers sleeping under a tree. 2 tigers wake up. The answer is 4 tigers. What is the question?

_____

_____

_____

_____

112   Chapter 2 • Lesson 1

Name ....

My Homework

Homework Helper

eHelp

Need help? connectED.mcgraw-hill.com

There are 4 rabbits playing together.
2 rabbits hop away.

How many rabbits are left playing near the carrots?

2 rabbits

Practice

Tell a number story. Use pennies to model if needed.
Write how many are left.

I.

How many birds are left on the window? _____ birds

Tell a number story. Use pennies to model if needed. Write how many are left.

2.

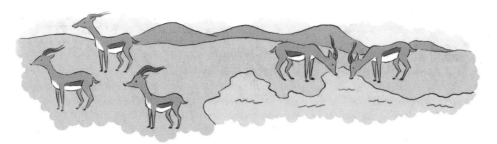

How many animals are still drinking water?

_____ animals

3.

How many animals are left standing still?

_____ animals

Brain Builders

4. **Test Practice** There are 8 hippos in a pond. 7 of the hippos get out. How many hippos are left in the pond? Draw a picture on a separate piece of paper to solve.

3 hippos ◯   2 hippos ◯   1 hippo ◯   0 hippos ◯

Math at Home Tell subtraction stories to your child. Have your child use objects such as stuffed animals, toy cars, or crayons to model each story.

Name

Lesson 2
Model Subtraction

ESSENTIAL QUESTION
How do you subtract numbers?

👐 Math in My World [Watch] [Tools]

⬤ Part	⚪ Part
8	_____
Whole	
10	

 Teacher Directions: Use ⬤⚪ to model. There are 10 toys in a toy box. Marty takes 8 toys out of the toy box. How many toys are left? Write the number.

Online Content at ⟋ **connectED.mcgraw-hill.com**   Chapter 2 • Lesson 2

Guided Practice

When you know the whole and one part, you can **subtract** to find the other part.

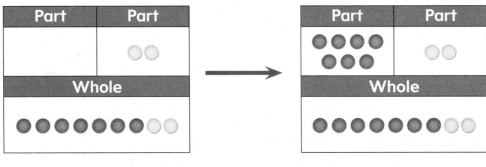

●Part	●Part
_____	2
Whole	
9	

→

●Part	●Part
7	2
Whole	
9	

Use Work Mat 3 and ●○ to subtract.

1.

●Part	●Part
4	_____
Whole	
5	

2.

●Part	●Part
_____	6
Whole	
8	

Talk Math You have 10 counters. 3 are yellow. Tell how you would use the part-part-whole mat to find how many are red. Explain.

●Part	○Part
	_____
Whole	

Independent Practice

Use Work Mat 3 and ⬤◯ to subtract.

3.

◯ Part	◯ Part
3	_____
Whole	
8	

4.

◯ Part	◯ Part
_____	3
Whole	
7	

5.

◯ Part	◯ Part
4	_____
Whole	
9	

6.

◯ Part	◯ Part
_____	1
Whole	
4	

7.

◯ Part	◯ Part
4	_____
Whole	
6	

8.

◯ Part	◯ Part
_____	3
Whole	
5	

9.

◯ Part	◯ Part
_____	5
Whole	
10	

10.

◯ Part	◯ Part
1	_____
Whole	
7	

Problem Solving

Solve. Use Work Mat 3 and **if needed.**

11. Clara saw 6 eagles on a limb.
2 eagles flew away. How many
eagles were left on the limb?

_____ eagles

Brain Builders

12. There are 10 crocodiles in a pond. After
some get out of the pond, there are 8
left. How many crocodiles got out of
the pond?

_____ crocodiles

Write Math The whole is 10 and one of the parts is
10. What is the other part? Explain.

_____

_ _

_____

_ _

_____

_ _

Name _____

My Homework

Homework Helper   eHelp

Need help? connectED.mcgraw-hill.com

When you know the whole and one of the parts, you can subtract to find the other part.

Part	Part
(coins)	(coin)
Whole	
(coins)	

→

Part	Part
7	1
Whole	
8	

Practice

Use pennies to subtract. Write the number.

1.

Part	Part
1	____
Whole	
5	

2.

Part	Part
2	____
Whole	
10	

3.

Part	Part
1	____
Whole	
6	

4.

Part	Part
6	____
Whole	
9	

Use pennies to subtract. Write the number.

5.

Part	Part
6	___
Whole	
7	

6.

Part	Part
4	___
Whole	
8	

7. There are 7 monkeys hanging from a branch. 3 monkeys go away. How many monkeys are still on the branch?

_____ monkeys

Brain Builders

8. There are 9 apes eating bananas. I ape stops eating. How many apes are still eating bananas? Explain your answer to a family member or friend.

_____ apes

Vocabulary Check

Circle the correct answer.

sum   **subtract**

9. You know the whole and one part. You can _____ to find the other part.

Math at Home Have your child use small objects such as cereal or beans to model subtraction.

Name ....

ESSENTIAL QUESTION ?
How do you subtract numbers?

 Math in My World [Watch ▶] [Tools ⬡]

_____ − _____ = _____

Write your subtraction sentence here.

 Teacher Directions: Use ⬤◯ to model. There are 7 zebras playing in a field. 5 of the zebras get in the pond. Draw an X on the zebras that get in the pond. How many zebras are still playing in the field? Write the subtraction number sentence.

Guided Practice

You can write a subtraction number sentence.

See

Say   **5**   **minus**   **2**   equals   **3.**

Write   _5_   **–**   _2_   **=**   _3_

$5 - 2 = 3$ is a **subtraction number sentence**.

3 is the **difference**.

Write a subtraction number sentence.

1.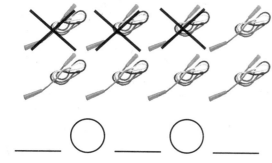

___ ◯ ___ ◯ ___

2.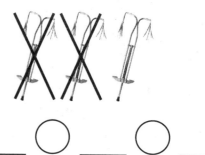

___ ◯ ___ ◯ ___

3.

___ ◯ ___ ◯ ___

4.

___ ◯ ___ ◯ ___

Talk Math   What does – mean?

Independent Practice

Write a subtraction number sentence.

5.

___ ◯ ___ ◯ ___

6.

___ ◯ ___ ◯ ___

7.

___ ◯ ___ ◯ ___

8.

___ ◯ ___ ◯ ___

9.

___ ◯ ___ ◯ ___

10.

___ ◯ ___ ◯ ___

11.

___ ◯ ___ ◯ ___

12.

___ ◯ ___ ◯ ___

13. There are 5 cars racing. 2 cars stop. How many cars are still racing?

_____ ◯ _____ ◯ _____ cars

Brain Builders

14. Kayla has 4 rockets. She gives some rockets away. She has 1 rocket left. How many rockets did she give away? Explain your thinking to a friend.

_____ ◯ _____ ◯ _____ rockets

15. By using two of these numbers each time, write as many subtraction number sentences as you can.

_____

- -

_____

- -

_____

- -

_____

Name ....

My Homework

Homework Helper

Need help? connectED.mcgraw-hill.com

You can write a subtraction number sentence.

$$8 - 2 = 6$$

Practice

Write a subtraction number sentence.

1.

____ ◯ ____ ◯ ____

2.

____ ◯ ____ ◯ ____

3.

____ ◯ ____ ◯ ____

4.

____ ◯ ____ ◯ ____

Write a subtraction number sentence.

5.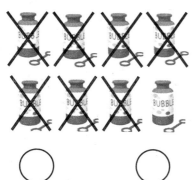

___ ◯ ___ ◯ ___

6.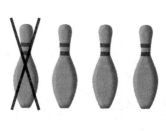

___ ◯ ___ ◯ ___

 **Brain Builders**

7. There are 7 elephants in a pond. Some elephants get out. Now there are 4. How many elephants got out of the pond?

_____ ◯ _____ ◯ _____ elephants

Is the number of elephants that are left more or fewer than 3 elephants?

_____

Vocabulary Check

Complete each sentence.

difference　　　　**subtraction number sentence**

8. 6 − 4 = 2 is a_____

_____.

9. In 3 − 2 = I, the _____ is I.

 Math at Home Using buttons, beans, or cereal, have your child act out subtraction stories. Have him or her write subtraction number sentences for the stories.

Name _____

ESSENTIAL QUESTION ?
How do you subtract numbers?

 Math in My World

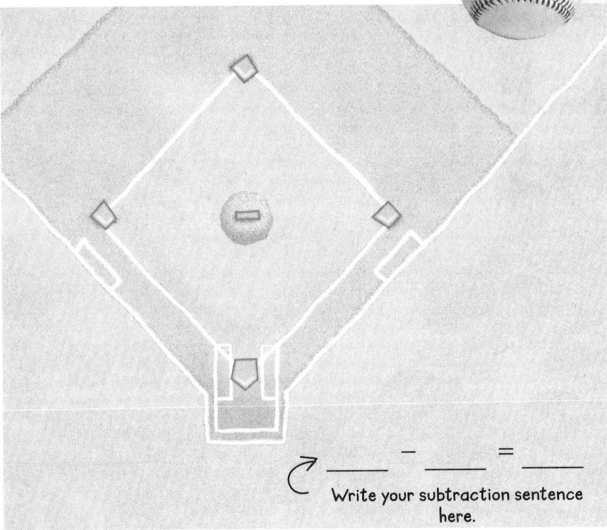

____ − ____ = ____

Write your subtraction sentence here.

 Teacher Directions: Use ●○ to model. A team had 6 baseballs at their game. They lost 6 of them. How many baseballs are left? Trace your counters. Mark Xs on the baseballs that are lost. Write the subtraction number sentence.

Guided Practice

When you subtract 0, you have the same number left.

 4 – 0 = _____ 4

When you subtract all, you have 0 left.

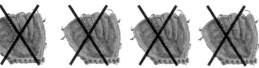 4 – 4 = _____ 0

Subtract.

1.

    5 – 5 = _____

2.

    8 – 0 = _____

3.

    1 – 0 = _____

4.

    3 – 3 = _____

Talk Math Why do you get zero when you subtract all? Explain.

Independent Practice

Subtract.

5.

   3 – 0 = _____

6.

   9 – 9 = _____

7.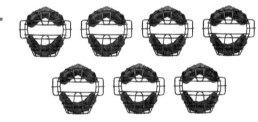

   7 – 0 = _____

8.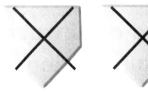

   2 – 2 = _____

9. 8 – 0 = _____

10. 3 – 3 = _____

11. 6 – 6 = _____

12. 4 – 0 = _____

13. 7 – 7 = _____

14. 1 – 1 = _____

15. 9 – 0 = _____

16. 5 – 0 = _____

Problem Solving

Write a subtraction number sentence.

17. There are 9 bears lying down.
0 of the bears leave. How many
bears are still lying down?

_____ – _____ = _____ bears

Brain Builders

18. Write your own subtract 0 or subtract
all subtraction number sentence. Tell a
friend a subtraction story that matches
the subtraction number sentence.

_____ – _____ = _____

19. A parrot has 6 baby birds in a nest.
0 baby birds fly away. The answer
is 6 baby birds. What is the question?

_____

_____

_____

_____

_____

Name _____

My Homework

Homework Helper

 eHelp

Need help? connectED.mcgraw-hill.com

When you subtract 0, you have the same number left.

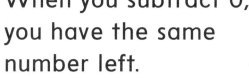

4 − 0 = 4

When you subtract all, you have 0 left.

4 − 4 = 0

Practice

Subtract.

1.

5 − 0 = _____

2.

6 − 6 = _____

3.

3 − 3 = _____

4.

7 − 0 = _____

Subtract.

5. $1 - 0 =$ _____

6. $9 - 9 =$ _____

7. $7 - 7 =$ _____

8. $6 - 0 =$ _____

9. $5 - 0 =$ _____

10. $8 - 8 =$ _____

11. There are 8 parrots on a branch. 0 parrots fly away. How many parrots are left on the branch?

_____ parrots

Brain Builders

12. **Test Practice** There are 9 lizards on a leaf. How many would have to crawl away so that there are no lizards on the leaf? Find the number sentence that matches the problem.

$9 + 9 = 18$ $9 - 0 = 9$ $9 - 9 = 1$ $9 - 9 = 0$

○ ○ ○ ○

Math at Home Give your child 3 objects. Have them use the objects to show $3 - 0$ and $3 - 3$.

Lesson 5
Vertical Subtraction

ESSENTIAL QUESTION
How do you subtract numbers?

Math in My World

Watch | Tools | Vocab

$$\begin{array}{r} \square \\ -\ \square \\ \hline \square \end{array}$$

$$\square - \square = \square$$

Teacher Directions: Use ⬤◯ to model. 4 bugs landed on each leaf. Then 3 bugs flew away from each of the leaves. How many bugs are left on each leaf? Trace the counters you used. Draw Xs on the counters to show the bugs that flew away. Write the subtraction number sentences.

Guided Practice

You can subtract across
or you can subtract down.
When the same numbers are
used, the answer is the same.

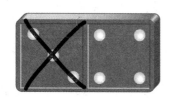

7 − 3 = ____4

$$\begin{array}{r} 7 \\ -\ 3 \\ \hline 4 \end{array}$$

Subtract.

1.

6 − 1 = ____

2.

5 − 1 = ____

3.
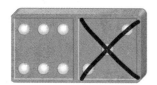

9 − 3 = ____

4.
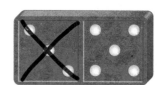

8 − 3 = ____

Talk Math How is subtracting down like
subtracting across?

Independent Practice

Subtract.

5.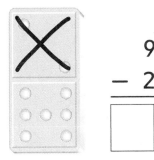

$$\begin{array}{r} 9 \\ -\ 2 \\ \hline \end{array}$$

6.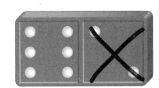

$8 - 2 =$ _____

7.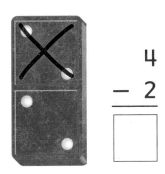

$$\begin{array}{r} 4 \\ -\ 2 \\ \hline \end{array}$$

8.

$6 - 2 =$ _____

9.

$7 - 5 =$ _____

10.

$$\begin{array}{r} 8 \\ -\ 6 \\ \hline \end{array}$$

11. $5 - 4 =$ _____

12. $8 - 5 =$ _____

13.

$$\begin{array}{r} 9 \\ -\ 1 \\ \hline \end{array}$$

14.

$$\begin{array}{r} 6 \\ -\ 6 \\ \hline \end{array}$$

Problem Solving

Write a subtraction number sentence.

15. There are 8 zebras eating grass.
2 zebras stop eating. How many
zebras are still eating grass?

_____ – _____ = _____ zebras

Brain Builders

16. There are 9 leopards in the field.
After some leopards leave,
there are 7 in the field. How
many left?

$$\begin{array}{r} \square \\ -\ \square \\ \hline \square \end{array}$$ leopards

Write Math   How is subtracting down different
from subtracting across?

_____

_____

_____

_____

_____

_____

Name

My Homework

Homework Helper Need help? connectED.mcgraw-hill.com

You can subtract across, or you can subtract down.

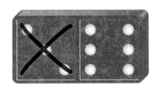

$9 - 3 = 6$

$$\begin{array}{r} 9 \\ -\ 3 \\ \hline 6 \end{array}$$

Practice

Subtract.

1.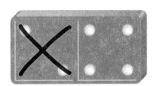

$6 - 2 =$ _____

2.

$$\begin{array}{r} 8 \\ -\ 6 \\ \hline \end{array}$$

3.

$$\begin{array}{r} 4 \\ -\ 2 \\ \hline \end{array}$$

4.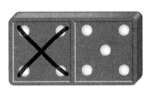

$9 - 4 =$ _____

Subtract.

5. $3 - 1 =$ _____

6. $9 - 6 =$ _____

7.
$$\begin{array}{r} 7 \\ -\ 2 \\ \hline \end{array}$$

8.
$$\begin{array}{r} 8 \\ -\ 1 \\ \hline \end{array}$$

9.
$$\begin{array}{r} 2 \\ -\ 2 \\ \hline \end{array}$$

10. There are 7 mangos growing on a tree. Diego picks 3 of the mangos. How many mangos are still on the tree?

_____ mangos

 Brain Builders

11. **Test Practice** The difference to a problem is 3. Which number sentence could have that difference?

$$\begin{array}{r} 10 \\ -\ 6 \\ \hline \end{array}$$
○

$$\begin{array}{r} 9 \\ -\ 6 \\ \hline \end{array}$$
○

$$\begin{array}{r} 7 \\ -\ 3 \\ \hline \end{array}$$
○

$$\begin{array}{r} 6 \\ -\ 2 \\ \hline \end{array}$$
○

 Math at Home Use 9 small objects. Show subtraction by taking some objects away. Have your child write the subtraction number sentence vertically and horizontally.

Name _____

Vocabulary Check

Draw lines to match.

1. subtract — —

2. minus sign — To take away a part from the whole.

3. subtraction number sentence — The answer in a subtraction problem.

4. difference — $6 - 5 = 1$

Concept Check

Tell a number story. Write how many are left.

5.

How many hippos are left in the water?

_____ hippos

Write a subtraction number sentence.

6.

_____ ◯ _____ ◯ _____

Subtract.

7.
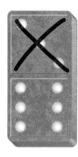
$$\begin{array}{r} 9 \\ -\ 3 \\ \hline \end{array}$$

8.

$7 - 2 =$ _____

---

9. There are 5 monkeys on a tree.
O monkeys get out of the tree. How
many monkeys are left on the tree?

_____ monkeys

Brain Builders

10. **Test Practice** The difference to the problem is 6.
Which number sentence has that difference?

◯ $5 + 2 =$

◯ $9 - 2 =$

◯ $8 - 2 =$

◯ $6 - 6 =$

Name _____

Lesson 6
Problem Solving
STRATEGY: Draw a Diagram

 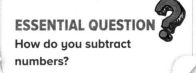
Abby has 5 erasers. She gives Matt 2 erasers. How many erasers does Abby have left?

1 Understand Underline what you know. Circle what you need to find.

2 Plan How will I solve the problem?

3 Solve I will draw a diagram.

_____3_____ erasers

4 Check Is my answer reasonable? Explain.

Lila has 8 toys. She lets Rex play with 3 of the toys. How many toys does Lila have left?

1 Understand Underline what you know.
Circle what you need to find.

2 Plan How will I solve the problem?

3 Solve I will...

_____ toys

4 Check Is my answer reasonable? Explain.

Name _____

Apply the Strategy

Draw a diagram to solve.

I. Nick has 6 cherries. He eats 3 of the cherries. How many cherries are left?

_____ cherries

Brain Builders

2. Alberto buys 7 apples. He eats 3 apples. How many apples does he have left? Explain how you solved the problem to a friend.

_____ apples

3. There are 7 oranges. Miles eats 5 oranges. How many oranges does Miles have left?

_____ oranges

Beyonce has 3 oranges. Who has more oranges, Miles or Beyonce?

_____

Choose a strategy
- Draw a diagram.
- Act it out.
- Write a number sentence.

4. Lena has 6 books. She gives 2 books to her sister. How many books does Lena have left?

_____ books

5. Jessica catches 4 frogs. 3 of the frogs hop away. How many frogs does Jessica have now?

_____ frog(s)

6. Marcos ate 5 crackers. Shani ate some of the crackers. Together they ate 9 crackers. How many crackers did Shani eat?

_____ crackers

Name _____

My Homework

Homework Helper eHelp   Need help? connectED.mcgraw-hill.com

Asia is watching 6 birds sitting
on a tree. 3 of the birds fly away.
How many birds are still on the tree?

1 Understand Underline what you know.
Circle what you need to find.

2 Plan How will I solve the problem?

3 Solve I will draw a diagram.

_____3_____ birds

4 Check Is my answer reasonable?

Problem Solving

Underline what you know. Circle what you need to find. Draw a diagram to solve.

1. There are 9 frogs on a tree.
   4 of the frogs hop away. How
   many frogs are left on the tree?

   _____ frogs

2. Max sees 7 butterflies on a
   flower. 5 of them fly away.
   How many butterflies are left?

   _____ butterflies

Brain Builders

3. There are 8 pandas sitting near a
   tree. 4 of them leave. How many
   pandas are still near the tree?
   Explain to a family member or
   friend how you know.

   _____ pandas

Math at Home Give your child a simple subtraction problem and have him or her solve it by drawing a picture.

Name _____

Lesson 7
Compare Groups

ESSENTIAL QUESTION
How do you subtract numbers?

 Math in My World

_____ monkeys

_____ birds

 Teacher Directions: Look at the picture. Place ⬛ on each bird.
Place ⬛ on each monkey. Count and write how many of each animal.
Circle the number that shows more.

Guided Practice

You can subtract to **compare** groups.

There are 6 zebras. There are 2 elephants. How many more zebras are there than elephants?

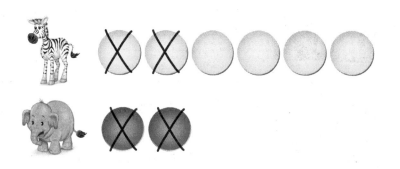

_6_ – _2_ = _4_ more zebras

_4_ fewer elephants

Use . Write a subtraction number sentence. Write how many more or fewer.

1. There are 7 giraffes. There are 2 rhinos. How many more giraffes are there than rhinos?

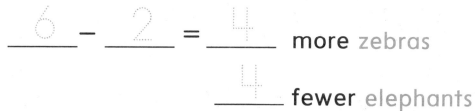

_____ – _____ = _____ more giraffes

Talk Math What happens when you compare equal groups?

Name _____

Use ◐ **. Write a subtraction number sentence.**
Write how many more or fewer.

2. There are 5 leopards. There are
   3 gorillas. How many fewer gorillas
   are there than leopards?

   _____ − _____ = _____ fewer gorillas

3. 9 butterflies are black. 5 butterflies
   are yellow. How many fewer butterflies
   are yellow than black?

   _____ − _____ = _____ fewer butterflies

4. There are 6 tigers. There are
   2 cheetahs. How many more tigers
   are there than cheetahs?

   _____ − _____ = _____ more tigers

Problem Solving

5. There are 8 elephants. There are 4 lions. How many fewer lions are there than elephants?

_____ fewer lions

Brain Builders

6. There are 9 monkeys and 8 parrots in a tree. How many more monkeys are there than parrots? Explain your answer to a friend.

_____ more monkey(s)

Write Math   How can you use subtraction to compare groups?

_____

_____

_____

_____

_____

Name _____

My Homework

Homework Helper

Need help? connectED.mcgraw-hill.com

There are 4 bears. There are 3 foxes. How many more bears are there than foxes?

$4 - 3 = 1$ more bear

Practice

Write a subtraction number sentence. Write how many more or fewer.

1. There are 4 snakes. There are 2 chimps. How many fewer chimps are there than snakes?

    _____ − _____ = _____ fewer chimps

2. Stella ate 3 bananas. Seth ate 2 bananas. How many fewer bananas did Seth eat than Stella?

    _____ − _____ = _____ fewer banana(s)

Write a subtraction number sentence.
Write how many more or fewer.

3. There are 5 frogs in a pond and 3 frogs on the land. How many more frogs are in a pond than on the land?

_____ − _____ = _____ more frogs

**Brain Builders**

4. Jesse caught 7 fish. Tia caught 6 fish. John caught 2 fish. How many fewer fish did Tia catch than Jesse?

_____ − _____ = _____ fewer fish

Vocabulary Check

Circle the correct answer.

compare   **add**

5. Subtract two different groups so you can _____ them to find which group has more and which has fewer.

 Math at Home Take a walk outside. Collect leaves or other objects found in nature. Create two groups with the items showing less than 9 items in each group. Have your child tell how many more or how many fewer.

Name _____

Copyright © McGraw-Hill Education   Ingram Publishing

Lesson 8
Subtract from 4 and 5

ESSENTIAL QUESTION
How do you subtract numbers?

Math in My World

Watch ▶   Tools

_____ − _____ = _____

Write your subtraction sentence here.

Teacher Directions: Use 🎲 to model. There are 4 crocodiles in a lake. 3 crocodiles get out. How many crocodiles are still in the lake? Trace your cubes and mark Xs on the cubes to show the crocodiles that get out. Write the subtraction number sentence.

Guided Practice

You can subtract from 4 and 5.

Subtract 2 from 4.

$4 - 2 = \underline{}$ The difference is 2.

Subtract 1 from 5.

$5 - 1 = \underline{}$ The difference is 4.

**Start with 4 . Subtract some cubes. Cross out .
Write different ways to subtract from 4.**

Subtract from 4

1. $4 - \underline{} = \underline{}$

2. $4 - \underline{} = \underline{}$

3. $4 - \underline{} = \underline{}$

4. $4 - \underline{} = \underline{}$

Talk Math   What does difference mean in subtraction?

Independent Practice

Start with 5 🔲. **Subtract some cubes. Cross out** 🔲.
Write different ways to subtract from 5.

Subtract from 5

5. 🔲🔲🔲🔲🔲   5 − _____ = _____

6. 🔲🔲🔲🔲🔲   5 − _____ = _____

7. 🔲🔲🔲🔲🔲   5 − _____ = _____

8. 🔲🔲🔲🔲🔲   5 − _____ = _____

9. 🔲🔲🔲🔲🔲   5 − _____ = _____

Subtract. Use Work Mat 3 and 🔲.

10. $4 - 1 =$ _____   11. $5 - 2 =$ _____

12. $5 - 5 =$ _____   13. $4 - 2 =$ _____

14. $\begin{array}{r} 5 \\ -\ 3 \\ \hline \end{array}$ 15. $\begin{array}{r} 5 \\ -\ 0 \\ \hline \end{array}$ 16. $\begin{array}{r} 4 \\ -\ 4 \\ \hline \end{array}$

Problem Solving

Write a subtraction number sentence.

17. Yuan draws 4 hippos. He crosses out 2.
How many hippos are there now?

_____ − _____ = _____ hippos

Brain Builders

18. Billy draws 5 lions. He crosses out 1.
How many lions are there now?
Explain how you solved the
problem to a friend.

_____ − _____ = _____ lions

19. Isabel wrote this subtraction
sentence. Tell why Isabel
is wrong. Correct it for her.

5 - 2 = 4

_____

_____

_____

_____

_____

Name _____

My Homework

Homework Helper

Need help? connectED.mcgraw-hill.com

You can subtract from 4 and 5.

4 − 1 = 3

5 − 3 = 2

Practice

Write different ways to subtract from 4 and 5.

1. 4 − _____ = _____

2. 4 − _____ = _____

3. 4 − _____ = _____

4. 4 − _____ = _____

5. 5 − _____ = _____

6. 5 − _____ = _____

7. 5 − _____ = _____

8. 5 − _____ = _____

Subtract.

9. 5 − 3 = _____

10. 4 − 4 = _____

Subtract.

11. $4 - 2 =$ _____

12. $5 - 1 =$ _____

13. $4 - 1 =$ _____

14. $5 - 4 =$ _____

15. $5 - 0 =$ _____

16. $4 - 3 =$ _____

17.
$$\begin{array}{r} 5 \\ -\ 5 \\ \hline \end{array}$$

18.
$$\begin{array}{r} 5 \\ -\ 2 \\ \hline \end{array}$$

19.
$$\begin{array}{r} 4 \\ -\ 0 \\ \hline \end{array}$$

20. Chad rents 5 movies. He watches 3 of the movies. How many movies does he have left to watch?

_____ movies

Brain Builders

21. **Test Practice** Which number sentence does not have a difference of 3?

○ $5 - 2 =$ _____   ○ $3 - 0 =$ _____

○ $4 - 1 =$ _____   ○ $5 - 3 =$ _____

 Math at Home Give your child 5 objects. Have him or her subtract different numbers from 4 or 5 and tell the difference.

Name ....

ESSENTIAL QUESTION
How do you subtract numbers?

 Math in My World Watch Tools

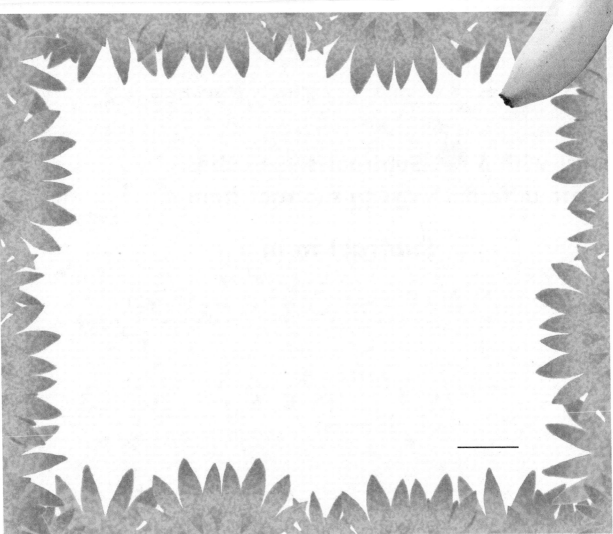

 Teacher Directions: Use 🎲 to model. There are 7 apes eating bananas. 3 apes stop eating bananas. How many apes are still eating bananas? Trace the cubes you used. Mark Xs on the cubes to show the apes that stop eating bananas. Write the number.

Guided Practice

You can subtract from 6 and 7.

Subtract 3 from 6.

$6 - 3 =$ _____ _3_   The difference is 3.

Subtract 5 from 7.

$7 - 5 =$ _____ _2_   The difference is 2.

Start with 6 ⬛. Subtract some cubes.
Write different ways to subtract from 6.

Subtract from 6

1. $6 -$ _____ $=$ _____   2. $6 -$ _____ $=$ _____

3. $6 -$ _____ $=$ _____   4. $6 -$ _____ $=$ _____

5. $6 -$ _____ $=$ _____   6. $6 -$ _____ $=$ _____

Talk Math   How could you use ⬛ to show subtraction?

Independent Practice

**Start with 7 ⬛. Subtract some cubes.
Write different ways to subtract from 7.**

Subtract from 7

7. 7 − _____ = _____   8. 7 − _____ = _____

9. 7 − _____ = _____   10. 7 − _____ = _____

11. 7 − _____ = _____   12. 7 − _____ = _____

Subtract. Use Work Mat 3 and ⬛.

13. 7 − 5 = _____   14. 6 − _____ = 4

15. 7 − _____ = 6   16. 6 − _____ = 0

17.   7
   − 0
   ─────

18.   7
   − 6
   ─────

19.   6
   − 5
   ─────

Problem Solving

Write a subtraction number sentence to solve.

20. 7 rhinos get a drink from a pond. 3 of them stop drinking. How many rhinos are still drinking?

_____ – _____ = _____ rhinos

 Brain Builders

21. Tell a friend a subtraction story. Write a subtraction sentence that matches the story.

Write Math What happens to the number of objects in a group when they are subtracted? Explain.

_____

_ _ _ _ _ _ _ _ _ _ _ _ _ _ _ _ _ _

_____

_ _ _ _ _ _ _ _ _ _ _ _ _ _ _ _ _ _

_____

_____

_ _ _ _ _ _ _ _ _ _ _ _ _ _ _ _ _ _

Name _____

My Homework

Homework Helper

Need help? connectED.mcgraw-hill.com

You can subtract from 6 and 7.

$$6 - 3 = 3$$

$$7 - 2 = 5$$

Practice

Write different ways to subtract from 6 and 7.

1. 6 − _____ = _____   2. 6 − _____ = _____

3. 6 − _____ = _____   4. 6 − _____ = _____

5. 7 − _____ = _____   6. 7 − _____ = _____

7. 7 − _____ = _____   8. 7 − _____ = _____

Subtract.

9. 6 − 1 = _____   10. 7 − 4 = _____

Subtract.

11. $7 - 1 =$ _____

12. $7 - 4 =$ _____

13. $6 - 5 =$ _____

14. $7 - 7 =$ _____

15. $\begin{array}{r} 7 \\ -\ 6 \\ \hline \end{array}$

16. $\begin{array}{r} 7 \\ -\ 0 \\ \hline \end{array}$

17. $\begin{array}{r} 6 \\ -\ 1 \\ \hline \end{array}$

18. There are 7 ants on a log. 5 of the ants crawl away. How many ants are left on the log?

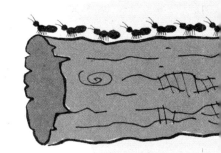

_____ ants

Brain Builders

19. **Test Practice** Ben sees 6 parrots on a branch. 2 of them fly away. How many parrots are left on the branch? Explain how you found your answer to a family member or friend.

2 parrots ○   4 parrots ○   5 parrots ○   8 parrots ○

 Math at Home Collect a group of 7 objects. Have your child show you how to subtract from 7. Have him or her write a subtraction number sentence.

Name ....

Check My Progress

Vocabulary Check

Circle the correct answer.

difference   **compare**

1. The words more and fewer can be used to
_____ the number of objects in two
different groups.

Concept Check

Write a subtraction number sentence.

2.

_____ – _____ = _____

3.

_____ – _____ = _____

Subtract.

4.   6
    – 4
   _____

5.   4
    – 4
   _____

6.   5
    – 1
   _____

Subtract.

7. $5 - 4 =$ _____

8. $6 - 0 =$ _____

9. $6 - 2 =$ _____

10. $7 - 7 =$ _____

11. $7 - 4 =$ _____

12. $4 - 2 =$ _____

13. There are 6 beetles. There are 4 butterflies. How many fewer butterflies are there than beetles?

_____ − _____ = _____ fewer butterflies

Brain Builders

14. **Test Practice** Which number sentence has the difference of 2?

$1 - 1 =$ _____
○

$6 - 3 =$ _____
○

$7 - 5 =$ _____
○

$5 - 4 =$ _____
○

Name _____

Lesson 10
Subtract from 8

ESSENTIAL QUESTION ❓
How do you subtract numbers?

 Math in My World Watch ▶ Tools

_____ − _____ = _____

↳Write your subtraction sentence here.

 Teacher Directions: Use 🎲 to model. 8 koalas are on a tree. 6 of the koalas climb away. How many koalas are still on the tree? Trace your cubes and mark Xs on the number of koalas that climb away. Write the subtraction number sentence.

Guided Practice

There are many ways to subtract from 8.

Subtract 4 from 8.

$8 - 4 = \underline{}$ The difference is 4.

Subtract 6 from 8.

$8 - 6 = \underline{}$ The difference is 2.

**Start with 8 . Subtract some cubes.
Write different ways to subtract from 8.**

Subtract from 8

1. $8 - \underline{} = \underline{}$ 2. $8 - \underline{} = \underline{}$

3. $8 - \underline{} = \underline{}$ 4. $8 - \underline{} = \underline{}$

5. $8 - \underline{} = \underline{}$ 6. $8 - \underline{} = \underline{}$

Talk Math   How do you know $8 - 5 = 3$? Explain.

Independent Practice

Use Work Mat 3 and **. Subtract.**

7. $7 - 3 =$ _____   8. $8 - 5 =$ _____

9. $8 - 1 =$ _____   10. $8 - 7 =$ _____

11. $8 - 0 =$ _____   12. $8 - 6 =$ _____

13. $8 - 2 =$ _____   14. $8 - 4 =$ _____

15. $8 - 8 =$ _____   16. $5 - 4 =$ _____

17.
$$\begin{array}{r} 8 \\ -\ 6 \\ \hline \end{array}$$

18.
$$\begin{array}{r} 6 \\ -\ 2 \\ \hline \end{array}$$

19.
$$\begin{array}{r} 7 \\ -\ 6 \\ \hline \end{array}$$

20.
$$\begin{array}{r} 8 \\ -\ 0 \\ \hline \end{array}$$

21.
$$\begin{array}{r} 6 \\ -\ 5 \\ \hline \end{array}$$

22.
$$\begin{array}{r} 8 \\ -\ 3 \\ \hline \end{array}$$

Problem Solving

Write a subtraction number sentence to solve.

23. There are 8 giraffes drinking water.
3 giraffes go away. How many
giraffes are left drinking water?

$$\underline{\hspace{1.5cm}} - \underline{\hspace{1.5cm}} = \underline{\hspace{1.5cm}}$$ giraffes

Brain Builders

24. Tell a friend a subtraction story. Write
a subtraction sentence that matches
the story.

25. Nia wrote this subtraction number
sentence. Tell why Nia is wrong.
Make it right.

$$\begin{array}{r} 8 \\ -\ 5 \\ \hline 2 \end{array}$$

_____

_ _

_____

_ _

_____

_ _

Name _____

My Homework

Homework Helper Need help? connectED.mcgraw-hill.com

There are many ways to subtract from 8.

$8 - 7 = 1$ $8 - 4 = 4$

Practice

Write different ways to subtract from 8.

1. 8 − _____ = _____   2. 8 − _____ = _____

3. 8 − _____ = _____   4. 8 − _____ = _____

5. 8 − _____ = _____   6. 8 − _____ = _____

7. 8 − _____ = _____   8. 8 − _____ = _____

Subtract.

9. $8 - 3 =$ _____   10. $8 - 4 =$ _____

Subtract.

11. $8 - 2 = $ _____

12. $7 - 4 = $ _____

13. $8 - 7 = $ _____

14. $5 - 5 = $ _____

15.
$$\begin{array}{r} 6 \\ -\ 3 \\ \hline \end{array}$$

16.
$$\begin{array}{r} 7 \\ -\ 5 \\ \hline \end{array}$$

17.
$$\begin{array}{r} 8 \\ -\ 8 \\ \hline \end{array}$$

18. FaShaun has 8 zebra stickers. He gives his friend 3 of them. How many stickers does FaShaun have left?

_____ stickers

 Brain Builders

19. **Test Practice** There are 8 rhinos drinking at a pond. 5 of the rhinos walk away. How many rhinos are still getting a drink? Draw a picture on a separate piece of paper to solve.

0 rhinos ○   2 rhinos ○   3 rhinos ○   14 rhinos ○

 Math at Home Collect a group of 8 objects. Have your child show you all of the numbers you can subtract from 8. Have him or her write each subtraction number sentence.

Name _____

Lesson 11
Subtract from 9

ESSENTIAL QUESTION
How do you subtract numbers?

 Math in My World Watch Tools

____ − ____ = ____

↪ Write your subtraction sentence here.

 Teacher Directions: Use 🎲 to model. There are 9 bugs on a flower. 2 of the bugs fly away. How many bugs are left? Trace your cubes and mark Xs on the bugs that fly away. Write the subtraction number sentence.

Guided Practice

There are many ways to subtract from 9.
Subtract 3 from 9.

$9 - 3 =$ _____ 6   The difference is 6.

Subtract 5 from 9.

$9 - 5 =$ _____ 4   The difference is 4.

**Start with 9 cubes. Subtract some cubes.
Write different ways to subtract from 9.**

Subtract from 9

1. $9 -$ _____ $=$ _____

2. $9 -$ _____ $=$ _____

3. $9 -$ _____ $=$ _____

4. $9 -$ _____ $=$ _____

5. $9 -$ _____ $=$ _____

6. $9 -$ _____ $=$ _____

7. $9 -$ _____ $=$ _____

8. $9 -$ _____ $=$ _____

Talk Math   How do you know when to subtract?

Independent Practice

Use Work Mat 3 and . Subtract.

9. $7 - 5 = $ _____

10. $9 - 3 = $ _____

11. $9 - 2 = $ _____

12. $5 - 3 = $ _____

13. $9 - 0 = $ _____

14. $9 - 5 = $ _____

15. $6 - 2 = $ _____

16. $9 - 4 = $ _____

17. $9 - 8 = $ _____

18. $7 - 7 = $ _____

19.
$$\begin{array}{r} 7 \\ -\ 0 \\ \hline \end{array}$$

20.
$$\begin{array}{r} 9 \\ -\ 6 \\ \hline \end{array}$$

21.
$$\begin{array}{r} 9 \\ -\ 9 \\ \hline \end{array}$$

22.
$$\begin{array}{r} 9 \\ -\ 7 \\ \hline \end{array}$$

23.
$$\begin{array}{r} 9 \\ -\ 1 \\ \hline \end{array}$$

24.
$$\begin{array}{r} 9 \\ -\ 2 \\ \hline \end{array}$$

Solve. Use Work Mat 3 and ▪ if needed.

25. Andre had 9 hats. He lost 5 of them.
How many hats does Andre have left?

_____ hats

⚙ Brain Builders

26. 7 ladybugs are climbing on a fence.
3 ladybugs fly away. How many
ladybugs are still on the fence? Explain
how you found your answer to a friend.

_____ ladybugs

Write Math   Why is the answer to a subtraction
problem called the difference?

_____

— — — — — — — — — — — — — — — — —

_____

— — — — — — — — — — — — — — — — —

_____

— — — — — — — — — — — — — — — — —

Name _____

My Homework

Homework Helper

Need help? connectED.mcgraw-hill.com

There are many ways to subtract from 9.

$$9 - 7 = 2$$

$$9 - 4 = 5$$

Practice

Write different ways to subtract from 9.

1. $9 -$ _____ $=$ _____

2. $9 -$ _____ $=$ _____

3. $9 -$ _____ $=$ _____

4. $9 -$ _____ $=$ _____

5. $9 -$ _____ $=$ _____

6. $9 -$ _____ $=$ _____

7. $9 -$ _____ $=$ _____

8. $9 -$ _____ $=$ _____

Subtract.

9. $9 - 3 =$ _____

10. $9 - 4 =$ _____

Subtract.

11. $9 - 2 =$ _____

12. $9 - 6 =$ _____

13. $7 - 3 =$ _____

14. $9 - 5 =$ _____

15.
$$\begin{array}{r} 9 \\ -\ 9 \\ \hline \end{array}$$

16.
$$\begin{array}{r} 5 \\ -\ 2 \\ \hline \end{array}$$

17.
$$\begin{array}{r} 9 \\ -\ 0 \\ \hline \end{array}$$

18. There are 9 frogs on a lily pad.
7 of the frogs hop in the water. How
many frogs are left on the lily pad?

_____ frogs

Brain Builders

19. **Test Practice** Which number sentence has
a difference of 4?

$7 - 2 =$ ____
◯

$8 - 5 =$ ____
◯

$9 - 4 =$ ____
◯

$9 - 5 =$ ____
◯

 Math at Home Collect a group of 9 objects. Have your child show you all of the
numbers they can subtract from 9. Have him or her write a subtraction
number sentence.

Name _____

Subtract from 10

ESSENTIAL QUESTION
How do you subtract numbers?

 Math in My World

_____ − _____ = _____

Write your subtraction sentence here.

 Teacher Directions: Use 🎲 to model. There are 10 tigers playing in a field. 4 tigers go away. How many tigers are left? Trace your cubes and mark Xs on the tigers that go away. Write the subtraction number sentence.

Guided Practice

There are many ways to subtract from 10.
Subtract 5 from 10.

10 − 5 = _____   The difference is 5.

Subtract 7 from 10.

10 − 7 = _____   The difference is 3.

Start with 10 **. Subtract some cubes.
Write different ways to subtract from 10.**

Subtract from 10

I. 10 − _____ = _____   2. 10 − _____ = _____

3. 10 − _____ = _____   4. 10 − _____ = _____

5. 10 − _____ = _____   6. 10 − _____ = _____

7. 10 − _____ = _____   8. 10 − _____ = _____

Talk Math   When would you use subtraction in
a real-world situation? Explain.

Independent Practice

Use Work Mat 3 and ▣. Subtract.

9. $10 - 3 = $ _____

10. $10 - 1 = $ _____

11. $9 - 5 = $ _____

12. $10 - 2 = $ _____

13. $10 - 0 = $ _____

14. $10 - 9 = $ _____

15. $6 - 3 = $ _____

16. $10 - 4 = $ _____

17. $10 - 8 = $ _____

18. $7 - 7 = $ _____

19. $10 - 7 = $ _____

20. $9 - 3 = $ _____

21.
$$\begin{array}{r} 10 \\ -\ 5 \\ \hline \end{array}$$

22.
$$\begin{array}{r} 10 \\ -\ 2 \\ \hline \end{array}$$

23.
$$\begin{array}{r} 4 \\ -\ 2 \\ \hline \end{array}$$

24.
$$\begin{array}{r} 8 \\ -\ 3 \\ \hline \end{array}$$

25.
$$\begin{array}{r} 10 \\ -\ 6 \\ \hline \end{array}$$

26.
$$\begin{array}{r} 10 \\ -\ 7 \\ \hline \end{array}$$

Problem Solving

Solve. Use Work Mat 3 and  if needed.

27. Marisa sees 10 flamingos in a lake. 2 of them get out of the lake. How many flamingos are still in the lake?

_____ flamingos

Brain Builders

28. There are some jeeps on a safari. 2 jeeps drove away. Then there were 7 jeeps. How many jeeps were at the safari to begin with?

_____ jeeps

29. There are 10 parrots on a branch. 7 of them fly away. The answer is 3 parrots. What is the question?

_____

_____

_____

_____

Name _____

My Homework

Homework Helper

Need help? connectED.mcgraw-hill.com

You can subtract many different numbers from 10.

10 − 9 = 1   The difference is 1.

10 − 5 = 5   The difference is 5.

Practice

Write different ways to subtract from 10.

1. 10 − _____ = _____   2. 10 − _____ = _____

3. 10 − _____ = _____   4. 10 − _____ = _____

5. 10 − _____ = _____   6. 10 − _____ = _____

7. 10 − _____ = _____   8. 10 − _____ = _____

Subtract.

9. $9 - 6 =$ _____

10. $10 - 4 =$ _____

11. $10 - 8 =$ _____

12. $10 - 5 =$ _____

13.
$$\begin{array}{r} 10 \\ -\ 9 \\ \hline \end{array}$$

14.
$$\begin{array}{r} 8 \\ -\ 2 \\ \hline \end{array}$$

15.
$$\begin{array}{r} 10 \\ -\ 0 \\ \hline \end{array}$$

Brain Builders

16. Chris saw some flies sitting on an elephant. After 6 flies flew away, 4 were left. How many flies were on the elephant to begin with?

_____ flies

17. **Test Practice** $10 - 2 =$ _____

6 $\bigcirc$ 7 $\bigcirc$ 8 $\bigcirc$ 9 $\bigcirc$

Math at Home Ask your child to use 10 buttons or pennies to show all of the ways they can subtract from 10. Have him or her write a subtraction number sentence to show one of the ways they subtracted from 10.

Name

Lesson 13
Relate Addition and Subtraction

ESSENTIAL QUESTION
How do you subtract numbers?

Math in My World Watch ▶ Tools

☐ + ☐ = ☐

☐ − ☐ = ☐

Write your number sentences here.

Teacher Directions: Use ●○ to model. There are 3 birds in the bird bath. 2 more birds join them. Write the addition number sentence to show how many birds there are in all. Write a related subtraction number sentence.

Online Content at 🔗 connectED.mcgraw-hill.com

Guided Practice

Related facts use the same numbers.
These facts can help you add and subtract.

$$5 + 2 = 7$$
$$7 - 5 = 2$$
$$7 - 2 = 5$$

Helpful Hint
$5 + 2 = 7$.
Use that fact to find
$7 - 2 = 5$.

You can use ___5___ + ___2___ = ___7___

to find ___7___ − ___2___ = ___5___.

They are opposite or inverse operations.

**Identify an addition fact. Use Work Mat 3 and ⬤◯
to find a related subtraction fact. Write both facts.**

1.

◯Part	◯Part
3	6
Whole	
9	

2.

◯Part	◯Part
2	4
Whole	
6	

____ + ____ = ____   ____ + ____ = ____

____ − ____ = ____   ____ − ____ = ____

Talk Math How can addition facts help you subtract? Explain.

Independent Practice

Use Work Mat 3 and ●○ to find the related subtraction facts. Write the facts.

3. $5 + 4 = 9$

_____ − _____ = _____

_____ − _____ = _____

4. $8 + 1 = 9$

_____ − _____ = _____

_____ − _____ = _____

5. $1 + 4 = 5$

_____ − _____ = _____

_____ − _____ = _____

6. $3 + 4 = 7$

_____ − _____ = _____

_____ − _____ = _____

7. $5 + 3 = 8$

_____ − _____ = _____

_____ − _____ = _____

8. $2 + 3 = 5$

_____ − _____ = _____

_____ − _____ = _____

Problem Solving

**Write a subtraction number sentence.
Then write a related addition fact.**

9. There are 8 lizards. 6 of the lizards
run away. How many lizards are left?

____ – ____ = ____ lizards

____ + ____ = ____

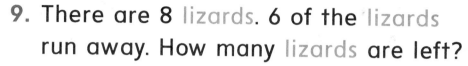

Brain Builders

10. Chad saw 7 birds. Audrey saw 4 birds.
How many more birds did Chad see?
Explain your thinking to a friend.

____ – ____ = ____ birds

____ + ____ = ____

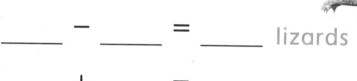 $5 + 4 = 9$ and $9 - 3 = 6$. Are these
related facts? Explain why or why not.

_____

_____

_____

_____

_____

Name _____

My Homework

Homework Helper

Need help? connectED.mcgraw-hill.com

You can write related addition and subtraction facts.
Related facts use the same numbers.

$$2 + 3 = 5$$
$$5 - 2 = 3$$
$$5 - 3 = 2$$

Practice

Identify an addition fact.
Write a related addition and subtraction fact.

1.

Part	Part
4	3
Whole	
7	

2.

Part	Part
3	5
Whole	
8	

____ + ____ = ____   ____ + ____ = ____

____ − ____ = ____   ____ − ____ = ____

Find the related subtraction facts. Write the facts.

3. $4 + 2 = 6$

____ – ____ = ____

____ – ____ = ____

4. $1 + 8 = 9$

____ – ____ = ____

____ – ____ = ____

Brain Builders

**Write the subtraction number sentence.
Then write the related addition fact.**

5. There are 6 flowers in a vase. 2 of them are pink, and the rest are yellow. How many yellow flowers are there?

____ – ____ = ____

____ + ____ = ____

Tell a family member or friend why knowing the subtraction fact helps you add.

Vocabulary Check

Circle the correct answer.

subtract   related facts

6. $3 + 1 = 4$ and $4 - 1 = 3$ are _____.

Math at Home Write an addition or subtraction number sentence using numbers 1–9. Have your child write a related addition or subtraction fact.

Name

Lesson 14
True and False Statements

ESSENTIAL QUESTION
How do you subtract numbers?

Math in My World Tools

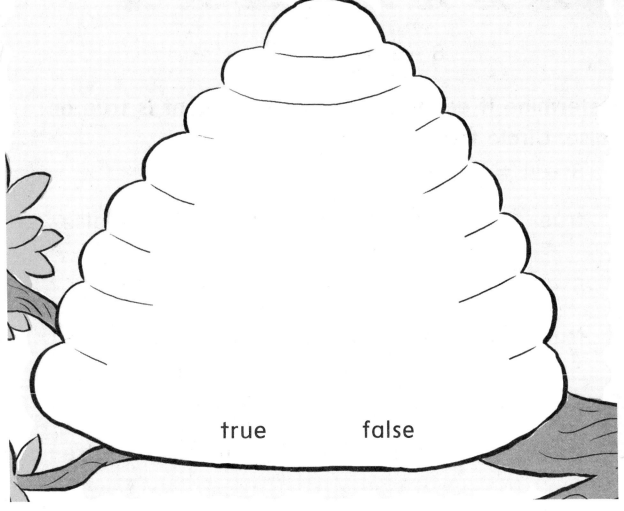

true   false

Teacher Directions: Use [cube] to model. There are 6 bees making honey. 3 bees fly away. Trace your cubes. Mark Xs to show the bees that fly away. Liam says that 4 bees are still making honey. Is this true or false? Circle it.

Guided Practice

In math, statements can be true or false.
A true statement is correct.

9 – 4 = 5 is true.

A false statement is incorrect.

8 – 3 = 1 is false.

Determine if each subtraction statement is true or false. Circle true or false.

1. 8 – 4 = 5

   true   false

2. 9 – 0 = 0

   true   false

3. 5 – 3 = 2

   true   false

4. 7 – 4 = 3

   true   false

5. 8 – 7 = 1

   true   false

6. 0 = 9 – 0

   true   false

Talk Math How do you know when a subtraction statement is true? Explain.

Independent Practice

Determine if each subtraction statement is true or false. Circle true or false.

7. $9 - 1 = 10$

true   false

8. $5 - 2 = 2$

true   false

9. $7 - 6 = 1$

true   false

10. $8 - 0 = 0$

true   false

11.
$$\begin{array}{r} 8 \\ -\ 2 \\ \hline 7 \end{array}$$
true

false

12.
$$\begin{array}{r} 4 \\ -\ 3 \\ \hline 1 \end{array}$$
true

false

13. $7 = 8 - 1$

true   false

14. $2 = 6 - 3$

true   false

15. $9 - 7 = 3$

true   false

16. $7 - 1 = 6$

true   false

Problem Solving

Determine if the word problem is true or false.
Circle true or false.

17. There are 7 spiders on a web.
    5 of those spiders crawl away.
    There are 2 spiders left on the web.

    true   false

 Brain Builders

18. Write a subtraction number sentence.
    Make up a true or false subtraction
    story about it. Ask a friend if it is true
    or false.

    _____ = _____ − _____

Write Math   Write your own false statement.
    Explain why your statement is false.

    _____

    _____

    _____

    _____

    _____

    _____

My Homework

Homework Helper Need help? connectED.mcgraw-hill.com

A true math statement is correct.
A false math statement is incorrect.

$$4 - 1 = 3$$ $$6 - 3 = 4$$

(true)   false   true   (false)

Practice

Determine if each subtraction statement is true or false. Circle true or false.

1. $8 - 8 = 8$

   true   false

2. $4 - 1 = 5$

   true   false

3. $6 = 9 - 3$

   true   false

4. $7 - 1 = 6$

   true   false

Determine if each subtraction statement is true or false. Circle true or false.

5.
$$\begin{array}{r} 6 \\ -\ 3 \\ \hline 3 \end{array}$$
    true

    false

6.
$$\begin{array}{r} 9 \\ -\ 1 \\ \hline 9 \end{array}$$
    true

    false

7. 5 birds are sitting on a branch.   true
   I bird flies away. There are
   6 birds left on the branch.   false

 Brain Builders

8. Can a number sentence be written like $9 - 4 = 5$
   or $5 = 9 - 4$?

_____

_____

_____

_____

9. **Test Practice** Which subtraction statement is true?

  $5 - 1 = 3$ $5 + 2 = 8$ $7 - 3 = 2$ $9 - 6 = 3$

    ○   ○   ○   ○

 Math at Home Write a false subtraction sentence using numbers 1–9. Ask your child if the problem is true or false. Have your child correct the number sentence.

Fluency Practice

Subtract.

1. 5 – 3 = _____

2. 10 – 4 = _____

3. 2 – 0 = _____

4. 6 – 3 = _____

5. 1 – 1 = _____

6. 9 – 5 = _____

7. 10 – 9 = _____

8. 7 – 3 = _____

9. 4 – 1 = _____

10. 6 – 5 = _____

11. 9 – 3 = _____

12. 8 – 0 = _____

13. 5 – 1 = _____

14. 10 – 3 = _____

15. 9 – 2 = _____

16. 4 – 4 = _____

17. 2 – 2 = _____

18. 5 – 2 = _____

19. 8 – 4 = _____

20. 7 – 6 = _____

21. 6 – 2 = _____

22. 9 – 1 = _____

23. 7 – 2 = _____

24. 4 – 2 = _____

Fluency Practice

Subtract.

1.   8
     − 3
    ⎯⎯

2.   10
     − 4
    ⎯⎯

3.   9
     − 6
    ⎯⎯

4.   5
     − 5
    ⎯⎯

5.   4
     − 2
    ⎯⎯

6.   6
     − 1
    ⎯⎯

7.   9
     − 0
    ⎯⎯

8.   3
     − 2
    ⎯⎯

9.   10
     − 5
    ⎯⎯

10.   7
     − 3
    ⎯⎯

11.   1
     − 1
    ⎯⎯

12.   8
     − 6
    ⎯⎯

13.   2
     − 0
    ⎯⎯

14.   9
     − 8
    ⎯⎯

15.   7
     − 2
    ⎯⎯

16.   5
     − 3
    ⎯⎯

17.   8
     − 4
    ⎯⎯

18.   6
     − 5
    ⎯⎯

19.   3
     − 3
    ⎯⎯

20. 10
     − 1
    ⎯⎯

21.   7
     − 5
    ⎯⎯

22.   9
     − 3
    ⎯⎯

23. 10
     − 3
    ⎯⎯

24.   4
     − 1
    ⎯⎯

Name _____

My Review

Vocabulary Check

Complete each sentence.

difference   related facts

subtract   subtraction number sentence

1. Addition and subtraction number sentences that use the same numbers are called _____.

2. 9 − 7 = 2 is a _____

   _____ .

3. When you take away, you _____.

4. In 7 − 4 = 3, the 3 is the _____.

Concepts Check

Write a subtraction number sentence.

5.

_____ ◯ _____ ◯ _____

Subtract.

6. $5 - 1 =$ _____

7. $4 - 2 =$ _____

8. $7 - 5 =$ _____

9. $9 - 5 =$ _____

10. $\begin{array}{r} 8 \\ -\ 4 \\ \hline \end{array}$

11. $\begin{array}{r} 6 \\ -\ 1 \\ \hline \end{array}$

12. $\begin{array}{r} 4 \\ -\ 0 \\ \hline \end{array}$

13. $\begin{array}{r} 9 \\ -\ 2 \\ \hline \end{array}$

14. $\begin{array}{r} 8 \\ -\ 3 \\ \hline \end{array}$

15. $\begin{array}{r} 7 \\ -\ 5 \\ \hline \end{array}$

Write the related subtraction facts.

16. $6 + 2 =$ _____

_____ $-$ _____ $=$ _____

_____ $-$ _____ $=$ _____

Determine if each subtraction statement is true or false. Circle true or false.

17. $10 - 3 = 7$

true   false

18. $3 = 6 - 2$

true   false

Name _____

Write a subtraction number sentence.

19. Angel has 7 bananas. She
    eats 2 of them. How many
    bananas are left?

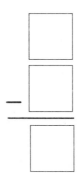

bananas

Circle true or false.

20. There are 9 birds on a tree. 3 of them
    fly away. 6 birds are still on the tree.

    true   false

Brain Builders

Find the matching subtraction number sentence.

21. **Test Practice** There are 5 elephants. There are
    3 leopards. How many fewer leopards are there?

    $5 - 3 = 1$ fewer leopard $5 - 3 = 2$ fewer leopards
           ○   ○

    $5 - 3 = 1$ fewer elephant $5 - 3 = 4$ fewer elephants
           ○   ○

Show ways to answer.

Find the missing part.

● Part	● Part
_____	3
Whole	
7	

Subtract 0.

8 − 0 = _____

ESSENTIAL QUESTION

How do you subtract numbers?

Subtract.

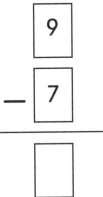

9
− 7
_____

Write a subtraction number sentence.

_____ − _____ = _____

Ready?
Set?
Solve!

Name _____ Date _____

Score _____

Performance Task

Brain Builders

At the Zoo

Mr. Russell works at the zoo. He takes care of the animals.

Show all your work to receive full credit.

Part A

In the morning, Mr. Russell feeds animals at the zoo. Write a subtraction number sentence to show how many animals he still needs to feed.

_____ − _____ = _____ animals

Part B

Mr. Russell had 9 bananas. The monkeys ate 4 of the bananas. How many bananas does Mr. Russell have left?

_____ − _____ = _____ bananas

Part C

Mr. Russell feeds 4 bears and 2 giraffes in the afternoon. How many animals does Mr. Russell have to feed in all?

_____ + _____ = _____ animals

Part D

Write a related subtraction sentence to Part C.

_____ - _____ = _____

Part E

Read the story. Check the number sentence. Circle *true* if it is correct. If not, circle *false*, and write the correct answer.

Mr. Russell had 8 bags of peanuts. He fed 3 bags of peanuts to the monkeys. How many bags of peanuts are left?

$$4 = 8 - 3$$

true   false

3 Addition Strategies to 20

ESSENTIAL QUESTION
How do I use strategies to add numbers?

We're in the Big City!

Watch a video!

Name _____

Chapter 3 Project

My Addition Story

1. Design a cover for your addition story book below. Title the book *My Addition Story Book.*

2. Write an addition story on each page. Include an addition number sentence and a drawing that matches each story.

3. Write at least 4 addition stories in all.

Name

_____

Am I Ready?

1. Circle the addition sign.

$$+ \quad - \quad =$$

2. Circle the equals sign.

$$+ \quad - \quad =$$

Add.

3. $\begin{array}{r} 3 \\ + \ 0 \\ \hline \end{array}$

4. $\begin{array}{r} 2 \\ + \ 2 \\ \hline \end{array}$

5. $\begin{array}{r} 7 \\ + 1 \\ \hline \end{array}$

6. $\begin{array}{r} 6 \\ + \ 3 \\ \hline \end{array}$

7. $\begin{array}{r} 4 \\ + \ 1 \\ \hline \end{array}$

8. $\begin{array}{r} 3 \\ + \ 4 \\ \hline \end{array}$

Use the pictures to write an addition number sentence.

9.

_____ ◯ _____ ◯ _____

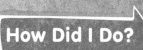

Shade the boxes to show the problems you answered correctly.

1	2	3	4	5	6	7	8	9

My Math Words

Review Vocabulary

minus (−)   plus (+)

Use the review words to fill in the blanks below.
Then write an addition and a subtraction
number sentence.

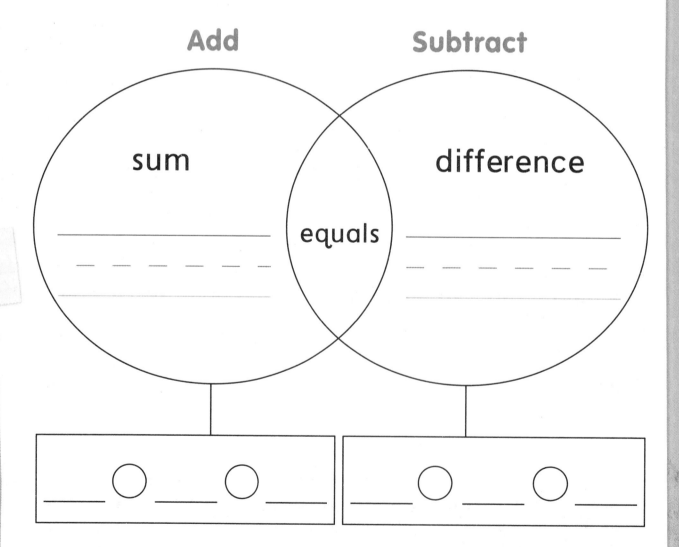

Add   **Subtract**

sum   difference

equals

My Vocabulary Cards

Processes & Practices

Lesson 3-4

addends

$$8 + 9 = 17$$

addends

Lesson 3-1

count on

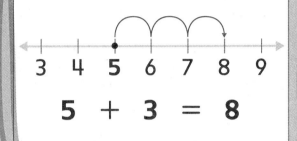

$$5 + 3 = 8$$

Lesson 3-4

doubles

$$3 + 3 = 6$$

Lesson 3-5

doubles minus 1

$$3 + 2 = 5$$

Lesson 3-5

doubles plus 1

$$3 + 4 = 7$$

Lesson 3-3

number line

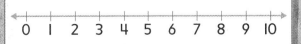

Teacher Directions:
Ideas for Use
• Have students arrange the cards in alphabetical order.

• Ask students to write a tally mark on each card every time they read or hear the word.

On a number line, start with the greater number and count up.

Any numbers being added together.

Add with doubles and subtract one.

Two addends that are the same.

A line with number labels.

Add with doubles and add one.

My Foldable

FOLDABLES® Follow the steps on the back to make your Foldable.

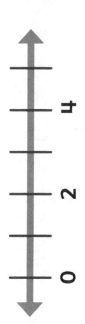

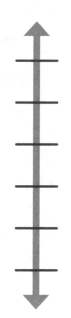

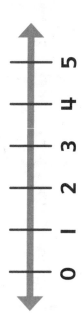

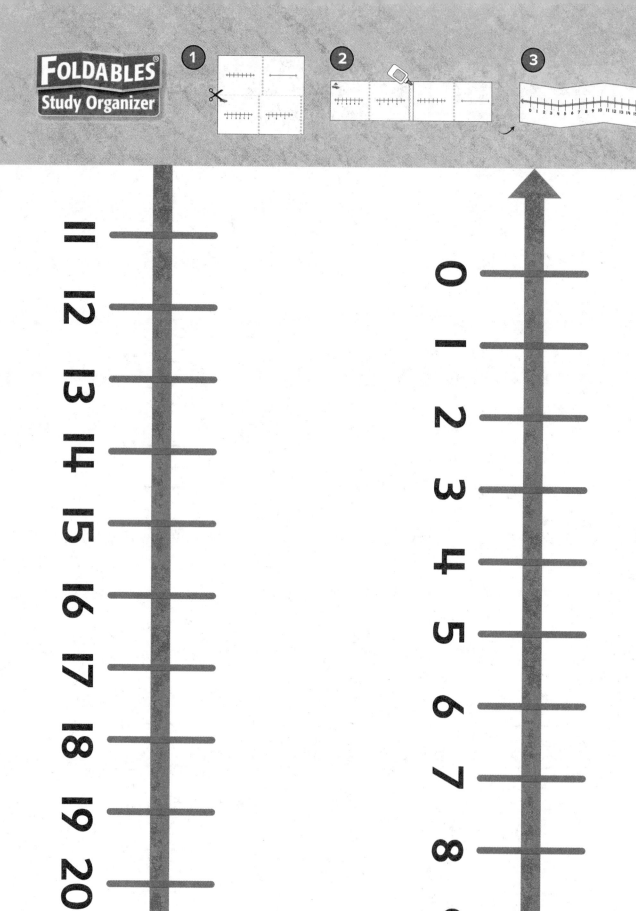

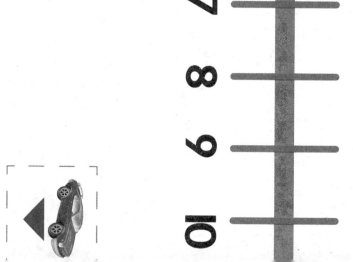

Name ....

Lesson 1
Count On 1, 2, or 3

ESSENTIAL QUESTION
How do I use strategies to add numbers?

 Math in My World Watch Tools

$$\text{\_\_\_\_} + \text{\_\_\_\_} = \text{\_\_\_\_}$$

Write your addition sentence here.

 Teacher Directions: Draw a group of four crayons. Then draw 1, 2, or 3 more crayons. Write an addition number sentence that tells how many in all.

Guided Practice

You can **count on** to add. There are 5 crayons in the group. Add 2 more crayons.

5, ___6___ , ___7___

$5 + 2 = $ ___7___

Helpful Hint
Start with the greater number, 5. Count on 2 more.
5, 6, 7
$5 + 2 = 7$

Use . Start with the greater number. Count on to add.

1.

7, _____ , _____ , _____

$7 + 3 = $ _____

Count on 3 is to add 3.

2.

6, _____ , _____

$2 + 6 = $ _____

Count on 2 is to add 2.

Talk Math

Tell how to count on to add $5 + 3$.

Name ....

Start with the greater number. Count on to add.

3. 5 + 3 = _____

4. 8 + 3 = _____

5. 4 + 1 = _____

6. 1 + 2 = _____

7. 9 + 3 = _____

8. 1 + 8 = _____

9. 3 + 7 = _____

10. 2 + 9 = _____

11. 1 + 7 = _____

12. 4 + 3 = _____

13. 2 + 7 = _____

14. 5 + 1 = _____

15.   8
     + 2
     ____

16.   1
     + 6
     ____

17.   9
     + 1
     ____

18.   3
     + 6
     ____

19.   2
     + 5
     ____

20.   3
     + 2
     ____

Problem Solving

21. Bella sees 2 buses. Then she sees 3 more buses. How many buses does she see in all?

_____ buses

Brain Builders

22. Jake saw 5 cars in the morning, 3 cars in the afternoon, and 3 more cars in the evening. How many cars did Jake see?

_____ cars

Write Math Explain how you count on to find 3 + 7. Which number do you start with?

_____

_____

_____

_____

_____

_____

Name _____

My Homework

Homework Helper Need help? connectED.mcgraw-hill.com

You can count on to add.

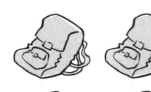

> **Helpful Hint**
> To count on, start with the greater number.

4,   5,   6,   7

$$4 + 3 = 7$$

Practice

Start with the greater number. Count on to add.

1. $7 + 2 =$ _____

2. $4 + 3 =$ _____

3. $5 + 3 =$ _____

4. $6 + 1 =$ _____

5. $9 + 1 =$ _____

6. $2 + 5 =$ _____

7. $8 + 3 =$ _____

8. $7 + 1 =$ _____

Start with the greater number. Count on to add.

9.
$$\begin{array}{r} 3 \\ + 6 \\ \hline \end{array}$$

10.
$$\begin{array}{r} 8 \\ + 2 \\ \hline \end{array}$$

11.
$$\begin{array}{r} 1 \\ + 8 \\ \hline \end{array}$$

12.
$$\begin{array}{r} 3 \\ + 9 \\ \hline \end{array}$$

13.
$$\begin{array}{r} 6 \\ + 2 \\ \hline \end{array}$$

14.
$$\begin{array}{r} 9 \\ + 2 \\ \hline \end{array}$$

Brain Builders

15. Drew went to basketball practice 5 times this week, soccer practice 3 times, and baseball practice 2 times. How many practices did Drew go to this week?

_____ practices

Explain what number you start with to count on to a family member or friend.

Vocabulary Check

Circle the missing word.

count on   sum

16. You can _____ to add when you join any number with 1, 2, or 3.

Math at Home Say a number between 1 and 9. Ask your child to add 1, 2, and 3 to that number.

Name

ESSENTIAL QUESTION
How do I use strategies to add numbers?

 Math in My World Tools

_____ pennies

Teacher Directions: Use ⬤ to model. Reese has 6 pennies in his bank. He puts in 3 more pennies. Count on to find how many pennies Reese has in all. Write how many there are in all. Trace the pennies to show your work.

Online Content at ✍ connectED.mcgraw-hill.com

Guided Practice

A penny has a value of 1 cent.
You can count on by ones to add pennies.

 or

penny
1 cent = 1¢

Helpful Hint
Start with 8 pennies.
Count on 9, 10, 11.
There are 11 pennies
in all.

8 ¢, _____9___ ¢, _____10___ ¢, _____11___ ¢

Count the group of pennies. Then count on to add.

1.

Count on 2
is to add 2.

7 ¢, _____ ¢, _____ ¢

Talk Math Why do you count on by ones when
you use pennies?

Independent Practice

Count the group of pennies. Then count on to add.

2.

4¢, _____¢, _____¢, _____¢

3.

10¢, _____¢, _____¢

4.

8¢, _____¢

Problem Solving

Processes &Practices

5. Kiah has 6 pennies. She is given 3 more. How many pennies does Kiah have now?

_____ pennies

Brain Builders

6. Enrique has 7 pennies in his left pocket. He has 2 more pennies in his right pocket. How many pennies does he have in all?

_____ pennies

At lunch Enrique got 3 more pennies. How many pennies does he have now?

_____ pennies

7. Eliana buys an eraser for 10 pennies. She buys a sticker for 2 pennies. The answer is 12 pennies. What is the question?

_____

- - - - - - - - - - - - - - - - - - - -

_____

- - - - - - - - - - - - - - - - - - - -

_____

- - - - - - - - - - - - - - - - - - - -

_____

Copyright © McGraw-Hill Education   ©Ken Karp/McGraw-Hill Education

Name ....

My Homework

Lesson 2

Count On Using Pennies

Homework Helper

Need help? connectED.mcgraw-hill.com

You can count on by ones to add pennies.

> **Helpful Hint**
> There are 8 pennies.
> Count on 9, 10. There
> are 10 pennies in all.

8¢,   9¢,   10¢

Practice

Count the group of pennies. Then count on to add.

I.

6¢,

_____¢, _____¢, _____¢

Copyright © McGraw-Hill Education   United States Mint image

Chapter 3 • Lesson 2   221

Count the group of pennies. Then count on to add.

2.

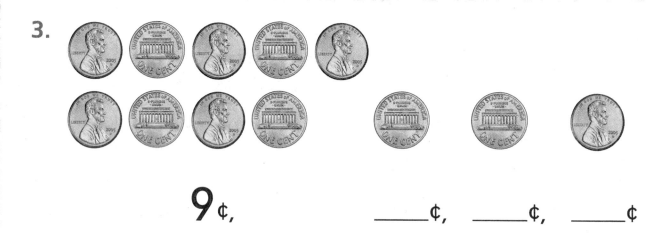

10 ¢,   _____ ¢

3.

9 ¢,   _____ ¢, _____ ¢, _____ ¢

Brain Builders

4. **Test Practice** Amal has 7 pennies. She also has 4 dimes. She is given 3 more pennies. How many pennies does Amal have in all?

9 pennies   10 pennies   11 pennies   14 pennies
   ○   ○   ○   ○

Math at Home Give your child 10 pennies. Provide your child with 2 more pennies. Ask him or her to count on by 2 using those pennies. Have your child tell you how many pennies he or she has in all.

Name _____

ESSENTIAL QUESTION
How do I use strategies to add numbers?

 Math in My World Watch Tools

0 1 2 3 4 5 6 7 8 9 10 11 12

_____ + _____ = _____

Write your addition sentence here.

 Teacher Directions: Use [cube] to model. A car is driving across the bridge. It starts at the number 4. It drives 3 spaces to the right. Where does the car stop? Write the addition number sentence.

Guided Practice

You can use a **number line** to add. Start with the greater number and count on by moving to the right.

Helpful Hint
Start with the greater number. 5, 6, 7, 8

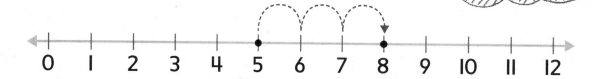

$$5 + 3 = \underline{8}$$

Use the number line to add. Show your work. Write the sum.

1. $6 + 2 = \underline{\hspace{2em}}$

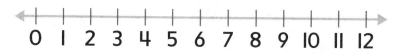

2. $8 + 2 = \underline{\hspace{2em}}$

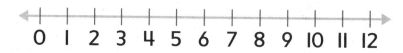

3. $1 + 4 = \underline{\hspace{2em}}$

4. $7 + 2 = \underline{\hspace{2em}}$

Talk Math How does a number line help you add?

Independent Practice

Use the number line to add. Write the sum.

5. 3 + 4 = _____   6. 2 + 9 = _____

7. 1 + 8 = _____   8. 7 + 3 = _____

9.   6   10.   8   11.   1
   + 1   + 3   + 9
   ____   ____   ____

12.   2   13.   9   14.   5
   + 7   + 3   + 2
   ____   ____   ____

15.   1   16.   4   17.   6
   + 7   + 2   + 3
   ____   ____   ____

Problem Solving

Use the number line to solve.

0 1 2 3 4 5 6 7 8 9 10 11 12

18. Amelio saw 5 bikes. His brother saw 2 bikes. How many bikes did they see in all?

_____ bikes

Brain Builders

19. Liz saw 8 pigeons. Logan saw 1 fewer pigeon than Liz. How many pigeons did they see in all?

_____ pigeons

Write Math   Why do you start with the greater number when adding on a number line?

_____

_____

_____

_____

_____

_____

Name _____

My Homework

Homework Helper

Need help? connectED.mcgraw-hill.com

You can use a number line to add.

$$7 + 3 = 10$$

Helpful Hint
Start with the greater number and count on by moving to the right.

Practice

Use the number line above to add. Write the sum.

1. $1 + 8 =$ _____

2. $8 + 2 =$ _____

3. $5 + 3 =$ _____

4. $8 + 3 =$ _____

5. $7 + 2 =$ _____

6. $4 + 1 =$ _____

7. $\begin{array}{r} 9 \\ + 2 \\ \hline \end{array}$

8. $\begin{array}{r} 4 \\ + 3 \\ \hline \end{array}$

9. $\begin{array}{r} 6 \\ + 3 \\ \hline \end{array}$

Use the number line to solve.

0 1 2 3 4 5 6 7 8 9 10 11 12

10. There are 6 people waiting for a taxi.
2 more people get in line. How many
people are waiting for a taxi in all?

_____ people

Brain Builders

11. Abigail puts 9 toy planes and 3 toy
cars on a shelf. How many toys did
she put on the shelf?

_____ toys

Explain to a family member or friend how you
used the number line to solve this problem.

Vocabulary Check

Complete each sentence.

number line   **count on**

12. When you add on a _____, start
with the greater number and move to the right.

13. To _____, start with the greater
number.

Math at Home Help your child create a number line from 0 to 10. Then, ask
your child to use the number line to show 1 + 9.

Name

Use Doubles to Add

 Math in My World Watch Tools

_____ + _____ = _____

Write your addition sentence here.

Teacher Directions: Use to model. Build a tower that has 3 red cubes and 3 green cubes. Trace your tower. Write the addition number sentence.

Guided Practice

Addends are the numbers you add.
Both addends are the same in a **doubles** fact.

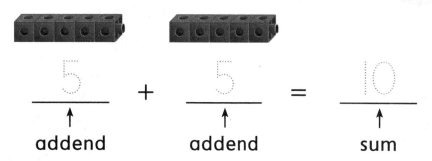

$$\underset{\underset{\text{addend}}{\uparrow}}{5} + \underset{\underset{\text{addend}}{\uparrow}}{5} = \underset{\underset{\text{sum}}{\uparrow}}{10}$$

Use 🎲 **to model. Complete the addition number sentence.**

1. ____ + ____ = ____

2. ____ + ____ = ____

Add. Circle the doubles facts.

3.   7
   + 7
   ———

4.   5
   + 5
   ———

5.   7
   + 4
   ———

6. 6 + 6 = ____

7. 9 + 9 = ____

Talk Math   Can you use doubles to make a sum of 7?

Name

_____

Independent Practice

Use to model. Complete the addition number sentence.

8.

_____ + _____ = _____

9.

_____ + _____ = _____

10.

_____ + _____ = _____

11.

_____ + _____ = _____

Add. Circle the doubles facts.

12.
$$\begin{array}{r} 8 \\ + 8 \\ \hline \end{array}$$

13.
$$\begin{array}{r} 9 \\ + 0 \\ \hline \end{array}$$

14.
$$\begin{array}{r} 9 \\ + 9 \\ \hline \end{array}$$

15. 8 + 3 = _____

16. 1 + 5 = _____

17. 6 + 6 = _____

18. 10 + 10 = _____

19. 3 + 3 = _____

20. 7 + 1 = _____

Problem Solving

21. 4 taxi cabs drove down the road. 4 more taxi cabs drove down the road. How many taxi cabs drove down the road in all?

_____ taxi cabs

Brain Builders

22. Emad saw 12 green and blue bicycles. The number of green bicycles is the same as the number of blue bicycles. Write a doubles fact to solve.

_____ = _____ + _____

Write Math   Is 3 + 6 a doubles fact? Explain.

_____

_____

_____

_____

_____

_____

_____

My Homework

Homework Helper

Need help? connectED.mcgraw-hill.com

In a doubles fact, both addends are the same.

$$3 \quad + \quad 3 \quad = \quad 6$$

↑   ↑   ↑

addend   addend   sum

Practice

Add. Circle the doubles facts.

1. 2
   + 2

2. 4
   + 4

3. 2
   + 9

4. 1
   + 6

5. 5
   + 5

6. 1
   + 1

7. $7 + 7 =$ _____

8. $7 + 2 =$ _____

Add. Circle the doubles facts.

9. 6 + 3 = _____

10. 4 + 4 = _____

11. There are 4 tan cats sitting on a fence.
4 black cats are also on the fence.
How many cats are on the fence in all?

_____ cats

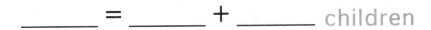

Brain Builders

12. There are 10 children playing hopscotch.
The same number of boys and girls are
playing. Write a number sentence to
show the number of boys and girls.

_____ = _____ + _____ children

Vocabulary Check

Complete each sentence.

doubles   **addends**

13. Numbers you add together to find a sum are

called _____.

14. Two addends that are the same number are

_____.

Math at Home Have your child identify things that show doubles such as
fingers on both hands, toes on both feet, or windows in a car.

Name _____

ESSENTIAL QUESTION ❓
How do I use strategies to add numbers?

 Math in My World

_____ + _____ = _____ ←

Write your addition sentence here.

 Teacher Directions: Use ⬛ to model. Show the doubles fact 3 + 3 on the billboard. Add or take away one cube from one of the groups of cubes. Trace the cubes. Write the addition number sentence.

Guided Practice

You can use near doubles facts to find a sum. If you know $5 + 5 = 10$, you can find $5 + 6$ and $5 + 4$.

doubles   **doubles plus I**   **doubles minus I**

      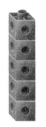

$5 + 5 = \underline{10}$ $5 + 6 = \underline{11}$ $5 + 4 = \underline{9}$

Use  to model. Write the addition number sentence.

I.

_____ + _____ = _____   _____ + _____ = _____

2.

_____ + _____ = _____   _____ + _____ = _____

Talk Math   How do doubles facts help you learn near doubles facts?

Name _____

Independent Practice

Use 🎲 to model. Find each sum.

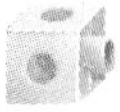

3. 2 + 2 = _____

  2 + 3 = _____

4. 3 + 3 = _____

  3 + 2 = _____

5. 8 + 7 = _____

6. 7 + 7 = _____

7. 6 + 6 = _____

8. 6 + 5 = _____

9.   7
   + 6
   ____

10.   7
   + 8
   ____

11.   4
   + 3
   ____

12.   5
   + 4
   ____

13.   5
   + 5
   ____

14.   5
   + 6
   ____

15.   7
   + 3
   ____

16.   7
   + 5
   ____

17.   7
   + 4
   ____

Problem Solving

Solve.

18. Tyra sees 5 beetles. Sara sees 6 beetles. How many beetles do they see in all? Write the doubles fact that helped you solve the problem.

_____ + _____ = _____   _____ beetles

 Brain Builders

19. Adam has 9 red flowers and 8 purple flowers. How many flowers does he have in all? Write two doubles facts that could help you solve the problem.

_____ flowers

_____ = _____ + _____   _____ = _____ + _____

20. Devon has 6 pets. Maya has 7 pets. The answer is 13 pets. What is the question?

_____

- - - - - - - - - - - - - - - - -

_____

- - - - - - - - - - - - - - - - -

_____

Name _____

My Homework

Homework Helper

Need help? connectED.mcgraw-hill.com

You can use near doubles facts to find a sum.

doubles	doubles plus 1	doubles minus 1
$3 + 3 = 6$	$3 + 4 = 7$	$3 + 2 = 5$

Practice

Find each sum.

1. $4 + 4 =$ _____

   $4 + 5 =$ _____

2. $8 + 8 =$ _____

   $8 + 7 =$ _____

3. $6 + 5 =$ _____

4. $6 + 6 =$ _____

5. $\begin{array}{r} 7 \\ + 6 \\ \hline \end{array}$

6. $\begin{array}{r} 7 \\ + 7 \\ \hline \end{array}$

7. $\begin{array}{r} 7 \\ + 8 \\ \hline \end{array}$

Solve. Write the doubles fact that helped you solve the problem.

8. Paul is a dog walker. He walks 7 small dogs. He walks 6 large dogs. How many dogs does he walk in all?

_____ dogs

_____ + _____ = _____

Brain Builders

Fill in the missing numbers to show a doubles minus one problem.

9. _____ girls walk to school. _____ boys walk to school. How many children walk to school in all?

_____ children

Vocabulary Check

Circle the missing vocabulary.

$$\text{doubles plus I} \qquad \text{doubles minus I}$$

10. If you know that $8 + 8 = 16$, then you can use _____ to find $8 + 7$.

Math at Home Give your child an addition problem such as $4 + 5$ or $3 + 4$. Have your child give you the doubles fact that will help him or her find the sum.

Name _____

Vocabulary Check

Draw lines to match.

1. **count on**

2. **number line**

3. **addends**

4. **doubles**

5. **doubles + 1**

6. **doubles − 1**

0 1 2 3 4 5 6

$4 + 4 = 8$

Start at a number and count forward to add.

Numbers you add together to find a sum.

Add with doubles and subtract one.

Add with doubles and add one.

Concept Check

Start with the greater number. Count on to add.

7.   6
     + 3
   _____

8.   1
     + 7
   _____

9.   2
     + 3
   _____

Use the number line to add. Write the sum.

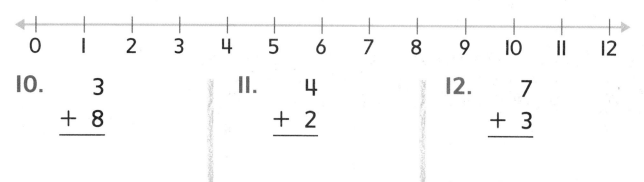

0 1 2 3 4 5 6 7 8 9 10 11 12

10.   3
    + 8
    ——

11.   4
    + 2
    ——

12.   7
    + 3
    ——

Add. Circle the doubles facts.

13. 4 + 4 = _____

14. 4 + 5 = _____

15. 9 + 8 = _____

16. 8 + 8 = _____

Brain Builders

17. Craig counted 7 doors. He counted 2 more doors. How many doors did Craig count in all? Draw a picture to solve the problem.

_____ doors

18. **Test Practice** Which doubles fact helps you find this sum?

9 + 8 = _____

10 + 10   7 + 7   5 + 5   9 + 9
   ○   ○   ○   ○

Lesson 6
Problem Solving
STRATEGY: Act It Out

ESSENTIAL QUESTION ❓
How do I use strategies
to add numbers?

3 red birds are on a branch.
There are 2 more yellow birds than
red birds on another branch.
How many yellow birds are there?

Watch ▶ Tools

1 **Understand** Underline what you know.
Circle what you need to find.

2 **Plan** How will I solve the problem?

3 **Solve** I will act it out.

_____5_____ yellow birds

4 **Check** Is my answer reasonable? Explain.

Susan found 3 shells on the beach. Jamar found I more shell than Susan did. How many shells did Jamar find?

1 Understand Underline what you know.
Circle what you need to find.

2 Plan How will I solve the problem?

3 Solve I will . . .

_____ shells

4 Check Is my answer reasonable? Explain.

Apply the Strategy

Act it out to solve.

1. Lou picked 7 apples. Jordan picked 1 more apple than Lou. How many apples did Jordan pick?

_____ apples

Brain Builders

2. The girls have 12 necklaces in all. Kim has the same number of necklaces as Lin. How many necklaces does each girl have?

_____ + _____ = _____   _____ necklaces

3. The clown sells toys in boxes of 2, 4, and 6. Nela's mom buys 2 boxes with 10 toys in all. Which two boxes does she buy? Explain how you solved the problem to a friend.

boxes with _____ and _____ toys

Choose a strategy
- Act it out.
- Draw a diagram.
- Write a number sentence.

4. There are 5 boats in the lake. There are 6 boats out of the lake. How many boats are there in all?

_____ boats

5. The girls ride in 2 taxis. The boys ride in some taxis. There are 9 taxis in all. How many taxis do the boys ride in?

⬤Part	⬤Part
2	
Whole	
9	

_____ taxis

6. Alan counts 7 lamps. Kurt counts the same number of lamps. How many lamps do Kurt and Alan count?

_____ lamps

Name _____

My Homework

Homework Helper

Need help? connectED.mcgraw-hill.com

There are 4 red pails. There are 3 more yellow pails than red pails. How many yellow pails are there?

1 Understand Underline what you know.
Circle what you need to find.

2 Plan How will I solve the problem?

3 Solve I will act it out.

There are 7 yellow pails.

4 Check Is my answer reasonable?

Problem Solving

Underline what you know. Circle what you need to find. Act out the problem to solve.

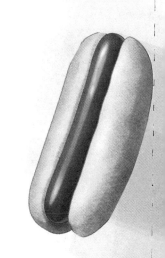

1. A hot dog vendor sold 9 hot dogs on Monday. She sold the same number of hot dogs on Tuesday. How many hot dogs did she sell in all?

_____ hot dogs

2. There are 4 people jogging. There are 5 more people walking than jogging. How many people are walking?

_____ people

Brain Builders

3. A monkey ate 5 peanuts. An elephant ate some peanuts. They ate 9 peanuts in all. How many peanuts did the elephant eat? Write a number sentence to solve.

_____ = _____ + _____

_____ peanuts

Math at Home Take advantage of problem-solving opportunities during daily routines such as riding in the car, doing laundry, putting away groceries, planning schedules, and so on.

Name _____

ESSENTIAL QUESTION ❓
How do I use strategies to add numbers?

✋ **Math in My World** ▶ Watch · Tools

Write your answer here.

_____

🦉 **Teacher Directions:** Use to model. There are 9 counters in the yellow ten-frame. There are 5 counters in the purple ten-frame. Move counters to make 10. Color the boxes used. Write how many counters there are in all.

Online Content at ⌖ connectED.mcgraw-hill.com

Guided Practice

You can make a 10 to help you add.

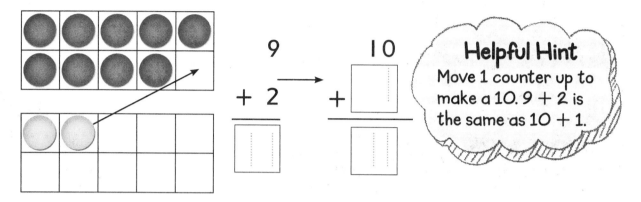

9
+ 2
———

10
+ ☐
———

Helpful Hint
Move 1 counter up to make a 10. 9 + 2 is the same as 10 + 1.

Use Work Mat 2 and ⬤◯. Make a ten to add.

1.

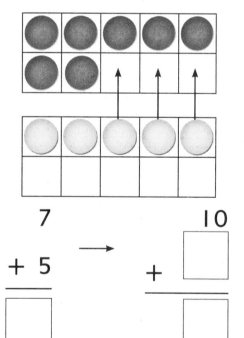

7
+ 5
———

10
+ ☐
———

2.

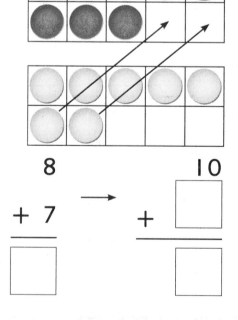

8
+ 7
———

10
+ ☐
———

Talk Math Why is it helpful to make a 10 on a ten-frame when finding sums greater than 10?

Independent Practice

Use Work Mat 2 and ⬤◯. Make a ten to add.

3.

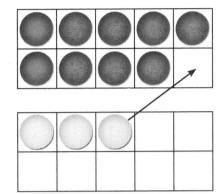

$$9$$
$$+\ 3$$
$$\overline{}\square$$

→

$$10$$
$$+\ \square$$
$$\overline{}\square$$

4.

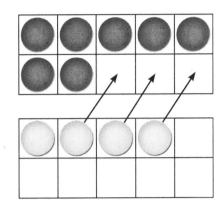

$$7$$
$$+\ 4$$
$$\overline{}\square$$

→

$$10$$
$$+\ \square$$
$$\overline{}\square$$

5.

$$9$$
$$+\ 6$$
$$\overline{}\square$$

→

$$10$$
$$+\ \square$$
$$\overline{}\square$$

6.

$$8$$
$$+\ 5$$
$$\overline{}\square$$

→

$$10$$
$$+\ \square$$
$$\overline{}\square$$

7.

$$8$$
$$+\ 4$$
$$\overline{}\square$$

→

$$10$$
$$+\ \square$$
$$\overline{}\square$$

8.

$$7$$
$$+\ 6$$
$$\overline{}\square$$

→

$$10$$
$$+\ \square$$
$$\overline{}\square$$

Problem Solving

Use Work Mat 2 and to solve.

9. Don has 8 goldfish. He gets 5 more.
   How many goldfish does he have now?

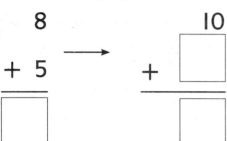

8

+ 5

□

10

+ □

□

_____ goldfish

Brain Builders

10. 6 pigs are in the mud. 5 more join
    them. How many pigs are in the mud?
    Tell a friend how you solved the problem.

_____ pigs

Write Math   Explain how to make a 10 to add
using ten-frames.

_____

_ _ _ _ _ _ _ _ _ _ _ _ _ _ _ _ _ _

_____

_ _ _ _ _ _ _ _ _ _ _ _ _ _ _ _ _ _

_____

_ _ _ _ _ _ _ _ _ _ _ _ _ _ _ _ _ _

Name

My Homework

Homework Helper Need help? connectED.mcgraw-hill.com

You can make a 10 to add.

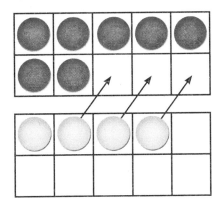

$$7 \longrightarrow 10$$
$$\underline{+\ 4} \qquad \underline{+\ 1}$$
$$11 \qquad\quad 11$$

Helpful Hint
Move 3 counters up to make a 10.

Make a ten to add.

1.

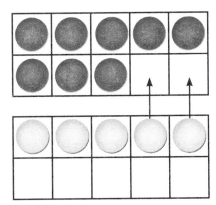

$$8 \longrightarrow 10$$
$$\underline{+\ 5} \qquad \underline{+\ }$$

2.

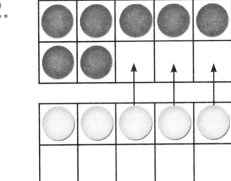

$$7 \longrightarrow 10$$
$$\underline{+\ 5} \qquad \underline{+\ }$$

Make a ten to add.

3.
$$
\begin{array}{r}
8 \\
+\ 9 \\
\hline
\square
\end{array}
\quad \longrightarrow \quad
\begin{array}{r}
10 \\
+\ \square \\
\hline
\square
\end{array}
$$

4.
$$
\begin{array}{r}
9 \\
+\ 4 \\
\hline
\square
\end{array}
\quad \longrightarrow \quad
\begin{array}{r}
10 \\
+\ \square \\
\hline
\square
\end{array}
$$

5.
$$
\begin{array}{r}
7 \\
+\ 8 \\
\hline
\square
\end{array}
\quad \longrightarrow \quad
\begin{array}{r}
10 \\
+\ \square \\
\hline
\square
\end{array}
$$

6.
$$
\begin{array}{r}
8 \\
+\ 6 \\
\hline
\square
\end{array}
\quad \longrightarrow \quad
\begin{array}{r}
10 \\
+\ \square \\
\hline
\square
\end{array}
$$

Brain Builders

7. There are 7 squirrels in the park. 5 more squirrels come to the park. How many squirrels are in the park now?

$$
\begin{array}{r}
7 \\
+\ 5 \\
\hline
\square
\end{array}
\quad \longrightarrow \quad
\begin{array}{r}
10 \\
+\ \square \\
\hline
\square
\end{array}
$$

_____ squirrels

Explain how you solved the problem to a family member or friend.

8. **Test Practice** Find 9 + 7.

14 ◯   15 ◯   16 ◯   17 ◯

Math at Home Draw two ten-frames. Give your child addition problems with sums to 20 and pennies. Help him or her solve the problems using the ten-frames and pennies.

Name _____

Lesson 8
Add in Any Order

 Math in My World Watch | Tools

$4 + 3 =$ _____ $3 + 4 =$ _____

 Teacher Directions: Use ●○ to model. Show 4 + 3. Write the sum. Now change the order. Trace and color to match. Write the sum. Describe what you notice about the sums.

Online Content at connectED.mcgraw-hill.com   **Chapter 3 • Lesson 8**   255

Guided Practice

You can change the order of the addends
and get the same sum.

$$\underline{3} + \underline{6} = \underline{9}$$

addend addend   sum

$$\underline{6} + \underline{3} = \underline{9}$$

addend addend   sum

Write the addends. Use **to add. Write the sum.**

1.

$$\underline{\qquad} + \underline{\qquad} = \underline{\qquad}$$

$$\underline{\qquad} + \underline{\qquad} = \underline{\qquad}$$

2. $\square$ $\square$

$+$ $\square$ $+$ $\square$

$\square$ $\square$

Talk Math Tell how you can show that $1 + 9$
has the same sum as $9 + 1$.

Independent Practice

Write the addends. Add. Write the sum.

3.

_____ + _____ = _____

_____ + _____ = _____

4.

_____ + _____ = _____

_____ + _____ = _____

5.

_____ + _____ = _____

_____ + _____ = _____

6.

_____ + _____ = _____

_____ + _____ = _____

Add.

7.   6   1   1
    + 1   + 7   + 6

8.   3   5   3
    + 3   + 3   + 5

Problem Solving

Write two ways to add. Solve.

9. 4 ladybugs climb onto some leaves. 8 more join them. How many ladybugs are on the leaves?

_____ + _____

_____ + _____   _____ ladybugs

Brain Builders

10. 3 butterflies are in the garden. 0 butterflies join them. How many butterflies are in the garden?

_____ + _____

_____ + _____   _____ butterflies

Explain how you solved the problem to a friend.

Write Math   Can you subtract in any order?
Use ◕. Explain.

_____

_ _

_____

_ _

_____

_ _

Name _____

Homework Helper

Need help? connectED.mcgraw-hill.com

You can change the order of the addends and get the same sum.

⬤⬤⬤⬤ ⚪⚪⚪⚪⚪ $4 + 5 = 9$

⚪⚪⚪⚪⚪ ⬤⬤⬤⬤ $5 + 4 = 9$

Practice

Write the addends. Then add.

1.

    ____ + ____ = ____

    ____ + ____ = ____

2.

    ____ + ____ = ____

    ____ + ____ = ____

Add. Circle the addends that have the same sum.

3. 6 + 5 = _____   5 + 6 = _____

4. 3 + 5 = _____   3 + 6 = _____

Brain Builders

Solve.

5. Kai walks up 7 steps first, and then walks up 8 more. Jermaine walks up 8 steps first, and then walks up 7 more. Who walked up more steps?

_____

_____

_____

6. **Test Practice** Lynn walks 3 blocks to the library. Then she walks 8 blocks to her dance lesson. How many blocks does she walk in all? Find the matching number sentence that solves this problem.

11 = 3 + 8   3 + 8 = 12   3 + 9 = 12   11 = 4 + 7
   ○   ○   ○   ○

Math at Home Show your child 4 plates and 2 cups. Have him or her write two addition sentences about them.

Name _____

 **Math in My World** Watch ▶ Tools

___ + ___ + ___ = ___

 Teacher Directions: Use ⬤◯ to model. 6 pets live in building A. 4 pets live in building B. 4 pets live in building C. Add the number of pets in two of the buildings. Then add the number of pets in the third. Write the addition number sentence.

Guided Practice

You can group numbers and add in any order. Look for doubles. Look for numbers that make a ten.

Doubles

$7 + 3 + 3$

6

Helpful Hint
3 + 3 is a double.

$7 + 6 = \underline{\hspace{1cm}}$

Make a 10

$7 + 3 + 3$

10

Helpful Hint
7 + 3 makes a 10.

$10 + 3 = \underline{\hspace{1cm}}$

Add the doubles or make a ten. Write that number. Add the other number to find the sum.

1. $6 + 4 + 3 = \underline{\hspace{1cm}}$

2. $4 + 7 + 4 = \underline{\hspace{1cm}}$

3. $\begin{array}{r} 3 \\ 3 \\ + \ 4 \\ \hline \end{array}$

4. $\begin{array}{r} 4 \\ 2 \\ + \ 4 \\ \hline \end{array}$

5. $\begin{array}{r} 4 \\ 6 \\ + \ 2 \\ \hline \end{array}$

Talk Math Tell how you would add the numbers $1 + 2 + 1$.

Independent Practice

Add the doubles or make a ten. Write that number.
Add the other number to find the sum.

6. ②+②+ 3 = _____

7. 4 +⑦+③= _____

8. 3 +⑨+① = _____

9. ④+④+ 2 = _____

10. ②+⑧+ 1 = _____

11. 4 +③+③= _____

12. ① ① + 8

13. ⑥ ④ + 2

Problem Solving

14. On Monday, 8 ducks were in the pond. On Tuesday and Friday 4 ducks were in the pond both days. How many ducks were in the pond in all?

____ + ____ + ____ = ____ ducks

Brain Builders

15. An apartment building has three floors. There are 6 doors on the first floor and 5 doors on the second floor. There are 15 doors in all. How many doors are on the third floor? Draw a picture to solve.

_____ doors

Write Math   Hayden says that 8 + 2 + 3 = 15. Tell why Hayden is wrong. Make it right.

_____

_ _ _ _ _ _ _ _ _ _ _ _ _ _ _ _ _

_____

_ _ _ _ _ _ _ _ _ _ _ _ _ _ _ _ _

_____

_ _ _ _ _ _ _ _ _ _ _ _ _ _ _ _ _

Name _____

My Homework

Homework Helper

Need help? connectED.mcgraw-hill.com

You can group numbers and add in any order. Look
for the doubles. Look for numbers that make a ten.

$6 + ④ + ④$

⬦ 8 ⬦

4 + 4 is a double.
4 + 4 = 8.
Then add 6 + 8.

$6 + 8 = 14$

$⑥ + ④ + 4$

⬦ 10 ⬦

6 + 4 makes a 10.
Then add 10 + 4

$10 + 4 = 14$

Practice

**Add the doubles or make a ten. Write that number.
Add the other number to find the sum.**

1. $⑤ + ⑤ + 2 =$ _____

   ⬦ ☐ ⬦

2. $⑦ + 4 + ③ =$ _____

   ⬦ ☐ ⬦

3. $② + 9 + ② =$ _____

   ⬦ ☐ ⬦

4. $③ + ⑦ + 4 =$ _____

   ⬦ ☐ ⬦

Write an addition number sentence to solve.

5. Bella sees 4 blue cars, 3 red cars, and 4 black cars. How many cars does she see in all?

_____ + _____ + _____ = _____ cars

6. Tom ate 1 slice of pizza. He also ate 9 carrots and 3 strawberries. How many pieces of food did he eat in all?

_____ + _____ + _____ = _____ pieces

7. Shannon buys 3 white shirts, 3 black shirts, and 4 green shirts. How many shirts does she buy in all?

_____ + _____ + _____ = _____ shirts

Brain Builders

8. **Test Practice** Find the number sentence that has a sum of 12.

$6 + 1 + 4 =$ _____   _____ $= 4 + 4 + 5$

       ○   ○

_____ $= 2 + 8 + 2$ $9 + 1 + 1 =$ _____

       ○   ○

Math at Home Place 3 crayons, 3 pencils, and 7 markers in front of your child. Have him or her identify how many there are of each and add the three numbers together to find how many there are in all. Have your child explain their work.

....

Fluency Practice

Add.

1. $4 + 4 =$ _____

2. $2 + 1 =$ _____

3. $1 + 1 =$ _____

4. $1 + 8 =$ _____

5. $1 + 0 =$ _____

6. $4 + 2 =$ _____

7. $6 + 3 =$ _____

8. $2 + 7 =$ _____

9. $3 + 2 =$ _____

10. $2 + 5 =$ _____

11. $6 + 1 =$ _____

12. $5 + 4 =$ _____

13. $\begin{array}{r} 7 \\ + 3 \\ \hline \end{array}$

14. $\begin{array}{r} 3 \\ + 3 \\ \hline \end{array}$

15. $\begin{array}{r} 2 \\ + 2 \\ \hline \end{array}$

16. $\begin{array}{r} 6 \\ + 4 \\ \hline \end{array}$

17. $\begin{array}{r} 0 \\ + 9 \\ \hline \end{array}$

18. $\begin{array}{r} 0 \\ + 5 \\ \hline \end{array}$

19. $\begin{array}{r} 7 \\ + 1 \\ \hline \end{array}$

20. $\begin{array}{r} 1 \\ + 3 \\ \hline \end{array}$

21. $\begin{array}{r} 0 \\ + 0 \\ \hline \end{array}$

Fluency Practice

Add.

1. 0 + 8 = _____

2. 3 + 5 = _____

3. 7 + 9 = _____

4. 6 + 6 = _____

5. 1 + 6 = _____

6. 4 + 3 = _____

7. 6 + 7 = _____

8. 7 + 7 = _____

9. 3 + 8 = _____

10. 2 + 3 = _____

11. 9 + 0 = _____

12. 8 + 1 = _____

13.
$$\begin{array}{r} 4 \\ + 7 \\ \hline \end{array}$$

14.
$$\begin{array}{r} 8 \\ + 4 \\ \hline \end{array}$$

15.
$$\begin{array}{r} 10 \\ + 2 \\ \hline \end{array}$$

16.
$$\begin{array}{r} 5 \\ + 2 \\ \hline \end{array}$$

17.
$$\begin{array}{r} 9 \\ + 6 \\ \hline \end{array}$$

18.
$$\begin{array}{r} 5 \\ + 6 \\ \hline \end{array}$$

19.
$$\begin{array}{r} 10 \\ + 10 \\ \hline \end{array}$$

20.
$$\begin{array}{r} 9 \\ + 9 \\ \hline \end{array}$$

21.
$$\begin{array}{r} 9 \\ + 8 \\ \hline \end{array}$$

Name ....

My Review

Vocabulary Check

Complete each sentence.

addends   count on   doubles

doubles minus 1   doubles plus 1   number line

1. Two addends that are the same number are called

   _____.

2. You can use a number line to _____
   numbers to find the sum.

3. Any numbers being added together are called

   _____.

4. To add with doubles plus one is called

   _____.

Concept Check

Start with the greater number. Count on to add.

5. 7 + 2 = _____   6. 1 + 6 = _____

7. 4 + 3 = _____   8. 5 + 5 = _____

Make a ten to add.

9.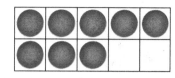

$$8 \quad\quad\quad 10$$
$$+\ 4 \quad\longrightarrow\quad +\ \boxed{}$$
$$\boxed{} \quad\quad\quad \boxed{}$$

10.
$$9 \quad\quad\quad 10$$
$$+\ 9 \quad\longrightarrow\quad +\ \boxed{}$$
$$\boxed{} \quad\quad\quad \boxed{}$$

11.
$$8 \quad\quad\quad 10$$
$$+\ 3 \quad\longrightarrow\quad +\ \boxed{}$$
$$\boxed{} \quad\quad\quad \boxed{}$$

Write the addends. Add.

12.

____ + ____ = ____

____ + ____ = ____

13.

____ + ____ = ____

____ + ____ = ____

Add the doubles or make a ten. Write that number.
Add the other number to find the sum.

14.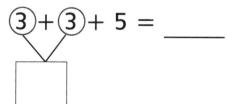

③ + ③ + 5 = ____

15. ① + 7 + ⑨ = ____

Name _____

 Problem Solving

16. Lea drank 2 glasses of milk.
Juan drank 3 glasses of milk.
How many glasses of milk did
they drink in all?

_____ glasses

 **Brain Builders**

17. Jenny has 7 books from the library.
Katy has I more book than Jenny. How
many books do they have in all?

_____ books

18. **Test Practice** 4 black dogs play at the park.
The same number of white dogs play there. How
many dogs are at the park?

4 dogs   6 dogs   8 dogs   10 dogs
   ○   ○   ○   ○

Show the strategies you use to add.

ESSENTIAL QUESTION

How do I use strategies to add numbers?

Count On

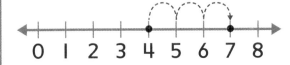

0 1 2 3 4 5 6 7 8

$4 + 3 =$ _____

Doubles and Near Doubles

$6 + 6 =$ _____

$6 + 7 =$ _____

$6 + 5 =$ _____

Add in Any Order

$8 + 1 =$ _____

$1 + 8 =$ _____

Add Three Numbers

$⑧ + 4 + ② =$ _____

[ ]

Now I know!!!

Performance Task

Brain Builders

The Soccer Game

The Comets and the Stars are playing soccer.

Part A

There are 9 players on the Comets that made it to the game on time. 2 players on the Comets team were late. Use the number line to show how many players there are on the Comets team all together. Write the number sentence.

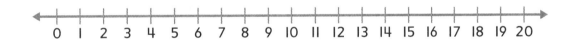

_____ + _____ = _____

Part B

In the first half of the game, the Comets made 7 goals. The Stars made 6 goals. How many goals were made in the first half of the game? Write a doubles fact that can help you solve the problem.

_____ goals

_____ + _____ = _____

Part C

During the game, first graders made 3 goals. Second graders made 9 goals and third graders made 7 goals. Write a number sentence to find how many goals were made in the game.

_____ + _____ + _____ = _____ goals

4 Subtraction Strategies to 20

ESSENTIAL QUESTION
What strategies can I use to subtract?

Let's Explore the Ocean!

Watch a video!

Watch

Name _____

Chapter 4 Project

My Story of the Chapter

1. Design a cover for your story of the chapter book below. Write a title for your book below.

2. Draw pictures, number sentences, or write vocabulary words that show the different things you learned in this chapter.

Name _____

Am I Ready?

1. Circle the minus sign.

$$+ \quad - \quad =$$

2. Circle the equals sign.

$$+ \quad - \quad =$$

Subtract.

3.   6
   − 0
   ____

4.   5
   − 2
   ____

5.   9
   − 8
   ____

6.   4
   − 3
   ____

7.   7
   − 3
   ____

8.   3
   − 2
   ____

9. Cross out 4 whales. Use the pictures to write a subtraction number sentence.

____ ◯ ____ ◯ ____

Shade the boxes to show the problems you answered correctly.

How Did I Do?

1	2	3	4	5	6	7	8	9

Online Content at connectED.mcgraw-hill.com

My Math Words

Vocab
abc

Review Vocabulary

| false | related facts | true |

Are the sets of number sentences related facts?
Circle true or false.

Number Sentences	True or False? Are they Related Facts?	
$3 + 4 = 7$ $7 - 3 = 4$	true	false
$5 + 2 = 7$ $9 - 2 = 7$	true	false
$8 + 4 = 12$ $12 - 4 = 8$	true	false

My Vocabulary Cards

Vocab

Lesson 4-1

count back

$$6 - 2 = 4$$

Lesson 4-7

fact family

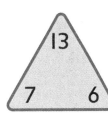

$$6 + 7 = 13$$
$$7 + 6 = 13$$
$$13 - 6 = 7$$
$$13 - 7 = 6$$

Lesson 4-8

missing addend

$$5 + \boxed{} = 9$$

$$9 - 5 = \boxed{}$$

- Have students draw examples for each vocabulary card. Have them make drawings that are different from what is shown on the card.

- Ask students to use the blank cards to write their own vocabulary cards.

Addition and subtraction sentences that use the same numbers.

On a number line, start at the greater number and count back.

You can use subtraction facts to find a missing addend. The missing addend is 4.

My Foldable

FOLDABLES® Follow the steps on the back to make your Foldable.

$12 - 4 = 8$

$15 - 6 = 9$

$14 - 7 = 7$

$11 - 9 = 2$

$17 - 9 = 8$

FOLDABLES® Study Organizer

1
12 − 4 = 8
15 − 6 = 9
14 − 7 = 7
11 − 9 = 2
17 − 9 = 8

2
12 − 4 = 8
15 − 6 = 9
14 − 7 = 7
11 − 9 = 2
17 − 9 = 8

3
4 + 8 = 12   8 + 4 = 12
15 − 6 = 9
14 − 7 = 7
11 − 9 = 2
17 − 9 = 8

$$\underline{4} + \underline{8} = \underline{12}$$

$$\underline{8} + \underline{4} = \underline{12}$$

$$\underline{} + \underline{} = \underline{}$$

$$\underline{} + \underline{} = \underline{}$$

$$\underline{} + \underline{} = \underline{}$$

$$\underline{} + \underline{} = \underline{}$$

$$\underline{} + \underline{} = \underline{}$$

$$\underline{} + \underline{} = \underline{}$$

$$\underline{} + \underline{} = \underline{}$$

$$\underline{} + \underline{} = \underline{}$$

Name _____

ESSENTIAL QUESTION ?
What strategies can I use to subtract?

 Math in My World Watch Tools

_____ squirrels

 Teacher Directions: Use ▪ to model. There are 5 squirrels playing. 3 of the squirrels run away. How many squirrels are still playing? Write the number.

Copyright © McGraw-Hill Education ©Steve Hamblin/Alamy

Processes
&Practices

Guided Practice

You can **count back** to subtract.

Start with 6. Count back 2.

6, 5, 4

6 − 2 = 4

Helpful Hint
Count back and take away cubes from the train. Write the numbers.

Count back to subtract. Use to help.

1. Start with 8.

8, ____, ____, ____

8 − 3 = ____

2. 7, ____

7 − 1 = ____

3. 4, ____, ____

4 − 2 = ____

4. 12, ____, ____

12 − 2 = ____

5. 10, ____, ____, ____

10 − 3 = ____

Talk Math How do you count back to find 7 − 2?

Independent Practice

Count back to subtract. Use ⬛ to help.

6. $11 - 2 =$ _____

7. $12 - 3 =$ _____

8. $3 - 2 =$ _____

9. $10 - 3 =$ _____

10. $9 - 3 =$ _____

11. $4 - 1 =$ _____

12. $6 - 1 =$ _____

13. $11 - 3 =$ _____

14. $7 - 3 =$ _____

15. $9 - 2 =$ _____

16.
$$\begin{array}{r} 10 \\ -\ 2 \\ \hline \end{array}$$

17.
$$\begin{array}{r} 9 \\ -\ 2 \\ \hline \end{array}$$

18.
$$\begin{array}{r} 8 \\ -\ 1 \\ \hline \end{array}$$

19.
$$\begin{array}{r} 12 \\ -\ 2 \\ \hline \end{array}$$

20.
$$\begin{array}{r} 10 \\ -\ 1 \\ \hline \end{array}$$

21.
$$\begin{array}{r} 5 \\ -\ 3 \\ \hline \end{array}$$

Problem Solving

Write a subtraction number sentence to solve.

22. There are 5 pelicans sitting on a rock. I of them flies away. How many pelicans are still sitting on the rock?

_____ − _____ = _____ pelicans

Brain Builders

23. There are II boats at the dock. Some leave, and then there are 3 boats at the dock. How many boats left the dock? Draw cubes to model your thinking.

_____ boats

WriteMath How do you count back to find II − 2? Explain.

_____

_____

_____

_____

Name _____

My Homework

Homework Helper

Need help? connectED.mcgraw-hill.com

Find 5 − 2. You can count back to subtract numbers.

Start with 5. Count back 2.

5,   4,   3   So, 5 − 2 = 3.

Practice

Count back to subtract.

1. 7, _____, _____

   7 − 2 = _____

2. 9, _____, _____, _____

   9 − 3 = _____

3. 12, _____, _____, _____

   12 − 3 = _____

4. 11, _____, _____

   11 − 2 = _____

5. 11 − 3 = _____

6. 8 − 1 = _____

Count back to subtract.

7. $5 - 2 =$ _____

8. $12 - 2 =$ _____

9. $10 - 3 =$ _____

10. $9 - 2 =$ _____

11. $\begin{array}{r} 8 \\ -\ 3 \\ \hline \end{array}$

12. $\begin{array}{r} 12 \\ -\ 1 \\ \hline \end{array}$

13. $\begin{array}{r} 11 \\ -\ 2 \\ \hline \end{array}$

Brain Builders

14. Landon had 9 masks. He lost some, but he still has 3 left. How many masks did Landon lose? Draw a picture to explain your thinking.

_____ masks

Vocabulary Check

Circle the correct answer.

count back   count on

15. Start at the number 6 and _____ by 2 to get the difference of 4.

 Math at Home Write $12 - 3 =$ ___. Have your child write the answer and explain how to count back to solve the problem.

x

x

x

286   Chapter 4 • Lesson 1

Name _____

 Math in My World [Watch ▶] [Tools]

Beach Balls for Sale

$$11 - 2 = \underline{\qquad}$$

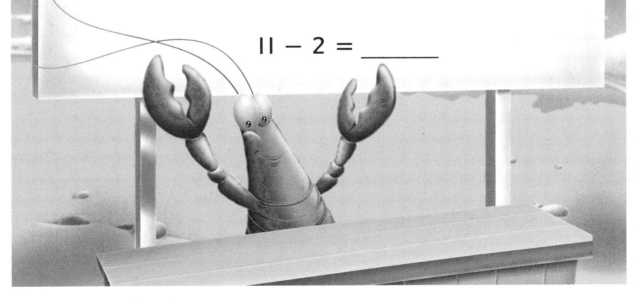

 Teacher Directions: Use a paper clip to count back on the number line. A store has 11 beach balls to sell. It sells 2 of them. How many beach balls do they have left? Write the number.

Guided Practice

You can use a number line to subtract.

9 − 3 = ___6___

Helpful Hint Start at 9. Count back 3 to find the difference. 9, 8, 7, 6.

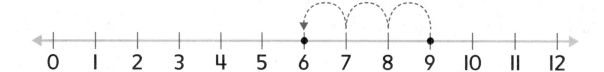

Use the number line to subtract. Show your work. Write the difference.

1. 8 − 2 = _____

2. 10 − 3 = _____

3. 5 − 1 = _____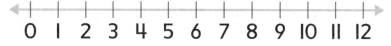

4. 12 − 3 = _____

Talk Math Can you only use the number line to help you subtract numbers? Explain.

Name _____

Use the number line to help you subtract.
Write the difference.

```
 0   1   2   3   4   5   6   7   8   9   10  11  12
```

5.   8
    − 3
    _____

6.   7
    − 3
    _____

7.   10
    − 2
    _____

8.   6
    − 2
    _____

9.   7
    − 1
    _____

10.   11
    − 3
    _____

11. $12 - 1 =$ _____

12. $9 - 2 =$ _____

13. $10 - 1 =$ _____

14. $11 - 2 =$ _____

15. $5 - 3 =$ _____

16. $10 - 3 =$ _____

17. $12 - 2 =$ _____

18. $7 - 2 =$ _____

Problem Solving

19. Ava sees 12 sharks. 2 of them swim away. How many sharks does Ava see now?

_____ sharks

Brain Builders

20. There are 11 manatees near the shore. 3 of them swim away. How many manatees are still near the shore? Draw a number line to show your thinking.

_____ manatees

Write Math How do you use a number line to help you subtract? Explain.

_____

_____

_____

_____

_____

_____

Name _____

My Homework

Homework Helper

Need help? connectED.mcgraw-hill.com

Use the number line to help you subtract.

|←+—+—+—+—+—+—+—+—+—+—+—+—+→|
0 1 2 3 4 5 6 7 8 9 10 11 12

$$7 - 3 = 4$$

Practice

Use the number line to subtract. Show your work.
Write the difference.

1. $10 - 3 = $ _____

2. $6 - 2 = $ _____

3. $12 - 3 = $ _____

4. $5 - 2 = $ _____

0 1 2 3 4 5 6 7 8 9 10 11 12

Use the number line to help you subtract.
Write the difference.

0 1 2 3 4 5 6 7 8 9 10 11 12

5. 11 − 2 = _____

6. 6 − 1 = _____

7. 9 − 3 = _____

8. 12 − 1 = _____

9. 6
    − 1
  ‾‾‾‾

10. 7
     − 3
   ‾‾‾‾

11. 8
     − 2
   ‾‾‾‾

12. There are 10 sharks swimming.
2 of the sharks swim away. How
many sharks are still swimming?

_____ sharks

Brain Builders

13. **Test Practice** What counting back problem has
the difference of 7?

○ 9 − 1 = _____

○ 10 − 2 = _____

○ 3 + 4 = _____

○ 10 − 3 = _____

Math at Home Have your child show 11 − 3 using a number line. Have him or
her explain how they are using the number line as they subtract.

Name _____

ESSENTIAL QUESTION
What strategies can I use to subtract?

Math in My World

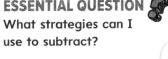 Watch Tools

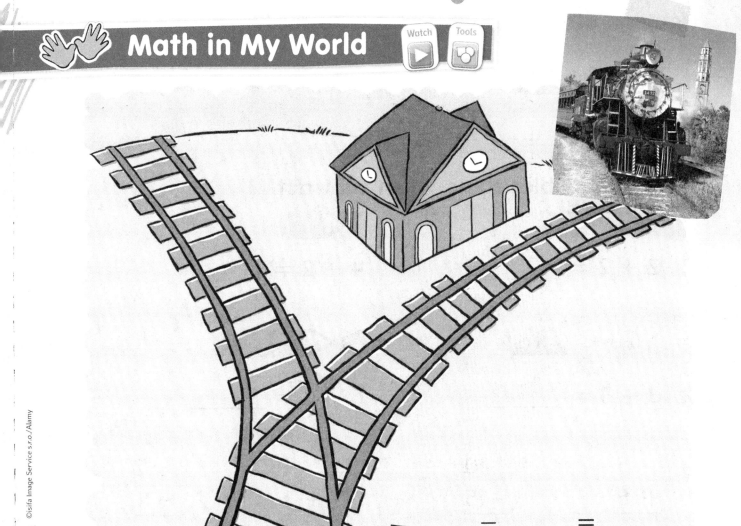

_____ − _____ = _____

Write your subtraction sentence here.

 Teacher Directions: Use 🎲 ⬛ to model. A train with 4 red train cars and 4 yellow train cars is on the tracks. The red train cars go to the right and the yellow train cars go to the left. Write the subtraction number sentence.

Guided Practice

You know how to use doubles facts to add.

4 + 4 = ___8___

You can also use doubles facts to subtract.

8 − 4 = ___4___

Add the doubles facts. Then subtract.

1. 2 + 2 = _____   4 − 2 = _____

2. 3 + 3 = _____   6 − 3 = _____

3.
$$\begin{array}{r} 5 \\ + 5 \\ \hline \end{array}$$
$$\begin{array}{r} 10 \\ - 5 \\ \hline \end{array}$$

4.
$$\begin{array}{r} 1 \\ + 1 \\ \hline \end{array}$$
$$\begin{array}{r} 2 \\ - 1 \\ \hline \end{array}$$

Talk Math   How can using doubles facts help you subtract?

Independent Practice

Add the doubles facts. Then subtract.

5. $10 + 10 = \underline{\hspace{1.5cm}}$

 $20 - 10 = \underline{\hspace{1.5cm}}$

6. $4 + 4 = \underline{\hspace{1.5cm}}$

 $8 - 4 = \underline{\hspace{1.5cm}}$

7.
$$\begin{array}{r} 8 \\ + 8 \\ \hline \end{array} \qquad \begin{array}{r} 16 \\ - 8 \\ \hline \end{array}$$

8.
$$\begin{array}{r} 6 \\ + 6 \\ \hline \end{array} \qquad \begin{array}{r} 12 \\ - 6 \\ \hline \end{array}$$

9.
$$\begin{array}{r} 5 \\ + 5 \\ \hline \end{array} \qquad \begin{array}{r} 10 \\ - 5 \\ \hline \end{array}$$

10.
$$\begin{array}{r} 9 \\ + 9 \\ \hline \end{array} \qquad \begin{array}{r} 18 \\ - 9 \\ \hline \end{array}$$

Add or subtract. Draw lines to match the doubles facts.

11. $10 + 10 = \underline{\hspace{1.5cm}}$

12. $9 + 9 = \underline{\hspace{1.5cm}}$

13. $8 + 8 = \underline{\hspace{1.5cm}}$

14. $7 + 7 = \underline{\hspace{1.5cm}}$

$14 - 7 = \underline{\hspace{1.5cm}}$

$16 - 8 = \underline{\hspace{1.5cm}}$

$18 - 9 = \underline{\hspace{1.5cm}}$

$20 - 10 = \underline{\hspace{1.5cm}}$

Problem Solving

15. Bella sees 18 crabs on the beach. 9 of the crabs go in the ocean. How many crabs does Bella now see on the beach?

_____ crabs

Brain Builders

16. Jose found 14 shells on the beach. He lost the same amount as he has left. Write the doubles addition fact Jose can use to find the number of shells Jose has left.

_____ = _____ + _____   _____ shells

17. Chase wrote 13 − 6 = 7 on the board to show a doubles fact. Tell why Chase is wrong. Make it right.

_____

_____

_____

_____

_____

_____

Name

My Homework

Homework Helper

Need help? connectED.mcgraw-hill.com

You can use doubles facts to help you subtract.

$$5 + 5 = 10$$ $$10 - 5 = 5$$

Practice

Add the doubles facts. Then subtract.

1. $4 + 4 =$ _____

   $8 - 4 =$ _____

2. $9 + 9 =$ _____

   $18 - 9 =$ _____

3. $8 + 8 =$ _____

   $16 - 8 =$ _____

4. $7 + 7 =$ _____

   $14 - 7 =$ _____

5. $3 + 3 =$ _____

   $6 - 3 =$ _____

6. $1 + 1 =$ _____

   $2 - 1 =$ _____

Add the doubles facts. Then subtract.

7. $\begin{array}{r} 5 \\ +\ 5 \\ \hline \end{array}$ $\begin{array}{r} 10 \\ -\ 5 \\ \hline \end{array}$ 8. $\begin{array}{r} 2 \\ +\ 2 \\ \hline \end{array}$ $\begin{array}{r} 4 \\ -\ 2 \\ \hline \end{array}$

9. $\begin{array}{r} 6 \\ +\ 6 \\ \hline \end{array}$ $\begin{array}{r} 12 \\ -\ 6 \\ \hline \end{array}$ 10. $\begin{array}{r} 10 \\ +10 \\ \hline \end{array}$ $\begin{array}{r} 20 \\ -10 \\ \hline \end{array}$

11. Camila has 6 pairs of sunglasses.
    She breaks 3 of them. How many pairs
    of sunglasses does she have left?

    _____ sunglasses

Brain Builders

12. **Test Practice** Find which number sentence is **not** a doubles fact.

    $16 - 8 = 8$ ⭘ $18 - 9 = 9$ ⭘ $7 + 7 = 14$ ⭘ $17 - 8 = 9$ ⭘

Math at Home Write a doubles fact such as 7 + 7 = 14. Ask your child to give you a related subtraction fact.

Name ....

Lesson 4
Problem Solving
STRATEGY: Write a Number Sentence

ESSENTIAL QUESTION
What strategies can I use to subtract?

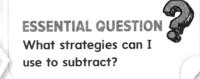

There are 6 seagulls flying over the ocean. 2 of the seagulls land in the ocean. How many seagulls are still flying?

1 **Understand** Underline what you know. Circle what you need to find.

2 **Plan** How will I solve the problem?

3 **Solve** I will write a number sentence.

____ ◯ ____ ◯ ____ seagulls

4 **Check** Is my answer reasonable? Explain.

There are 11 children building sand castles on the beach. 3 of the children go home. How many children are still building sand castles?

1 Understand Underline what you know. Circle what you need to find.

2 Plan How will I solve the problem?

3 Solve I will...

_____ ◯ _____ ◯ _____ children

4 Check Is my answer reasonable? Explain.

Apply the Strategy

Write a subtraction number sentence to solve.

1. Isaac bought 12 shells. He lost 6 of the shells. How many shells does Isaac have now?

_____ ◯ _____ ◯ _____ shells

2. Juan found 11 starfish on the beach. He gave 4 of them to his sister. How many starfish does Juan have now?

_____ ◯ _____ ◯ _____ starfish

Brain Builders

3. There are 12 sea horses swimming together. Some swam away, and now there are 8 sea horses. How many sea horses swam away?

_____ ◯ _____ ◯ _____ sea horses

Choose a strategy
- Write a number sentence.
- Draw a diagram.
- Act it out.

4. Brody saw 12 jellyfish in the ocean. Hunter saw 7 jellyfish there. How many more jellyfish did Brody see than Hunter?

_____ more jellyfish

5. There are 5 alligators swimming. 3 of them get out of the water. How many of the alligators are still swimming?

_____ alligators

6. Lina had 10 sandals. She gave away 6 of them. How many sandals does Lina have left?

_____ sandals

Name _____

My Homework

Homework Helper eHelp   Need help? connectED.mcgraw-hill.com

There are 12 dolphins swimming
together. 6 of the dolphins swim away.
How many dolphins are still swimming together?

1 Understand Underline what you know.
Circle what you need to
find.

2 Plan How will I solve the problem?

3 Solve I will write a number sentence.

$$12 - 6 = 6$$

6 dolphins are still swimming together.

4 Check Is my answer reasonable?

Problem Solving

Underline what you know. Circle what you need to find. Write a subtraction number sentence to solve.

1. There are 9 flamingos in the water. 5 flamingos get out. How many flamingos are still in the water?

_____ ◯ _____ ◯ _____ flamingos

2. 13 stingrays are swimming together. 4 of them swim away. How many stingrays are still swimming together?

_____ ◯ _____ ◯ _____ stingrays

Brain Builders

3. Sydney saw 11 hermit crabs on the beach. He saw 6 more hermit crabs than Noah. How many hermit crabs did Noah see?

_____ ◯ _____ ◯ _____ hermit crabs

Math at Home Give your child a subtraction problem about things around the house. Have him or her write a subtraction number sentence to solve the problem.

Name _____

Vocabulary Check

Circle the correct answer.

count back   count on

1. Start at a number and _____ to subtract.

Concept Check

Count back to subtract.

2. 9 − 1 = ____   3. 11 − 3 = ____

4. 5 − 2 = ____   5. 7 − 3 = ____

6. 10 − 1 = ____   7. 8 − 3 = ____

8.   12   9.   6   10.   11
  − 2   − 2   − 2

Use the number line to subtract. Write the difference.

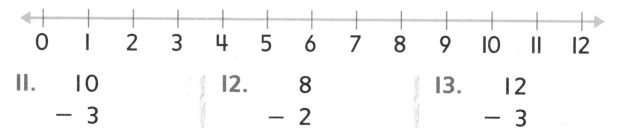

11.   10
     − 3
	12. 8
	     − 2
		13. 12
		     − 3

Add the doubles facts. Then subtract.

14.   4   8   15.   9   18
       + 4   − 4   + 9   − 9

Brain Builders

Draw a picture to solve.

16. There are 10 bikes parked at the beach. 2 children leave with their bikes. How many bikes are still at the beach?

_____ bikes

17. **Test Practice** There are 14 whales swimming together. 7 of them swim away. How many whales are still swimming together?

6 whales   7 whales   14 whales   21 whales

                        ○

Name _____

Lesson 5
Make 10 to Subtract

ESSENTIAL QUESTION ?
What strategies can I use to subtract?

 Math in My World

$$13 - 7 = $$

$$3 \qquad 4$$

$$13 - 3 = 10$$

$$10 - 4 = 6$$

 Teacher Directions: Use ▨ to model. There are 13 coconuts on the island. 7 of the coconuts roll into the ocean. How many coconuts are left on the island? Trace the numbers. Draw the number of coconuts that are left on the island.

Online Content at connectED.mcgraw-hill.com

Chapter 4 • Lesson 5   307

Copyright © McGraw-Hill Education ©Ingram Publishing/age fotostock

Guided Practice

To subtract, first take apart a number
to make a 10. Then subtract.

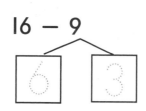

$16 - 9$

6   3

$16 - \underline{6} = 10$

$10 - \underline{3} = \boxed{7}$

Helpful Hint
Think $16 - 6 = 10$.
So, break apart
9 as 6 and 3.

$16 - 6 = 10$
and
$10 - 3 = 7$

So, $\underline{16} - \underline{9} = \underline{7}$.

Use 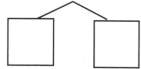 **and** . **Take apart the number
to make a 10. Then subtract.**

1. $12 - 7$

    ☐   ☐

    $12 - \underline{\quad} = \boxed{\quad}$

    $\underline{\quad} - \underline{\quad} = \boxed{\quad}$

    So, $12 - 7 = \underline{\quad}$.

2. $15 - 6$

    ☐   ☐

    $15 - \underline{\quad} = \boxed{\quad}$

    $\underline{\quad} - \underline{\quad} = \boxed{\quad}$

    So, $15 - 6 = \underline{\quad}$.

Talk Math   Explain how you can make a ten to
find $13 - 7$.

Name _____

Independent Practice

Use ▭▭▭▭▭ and ▪. Take apart the number
to make a 10. Then subtract.

3. 17 − 9

17 − _____ = ▢

_____ − _____ = ▢

So, 17 − 9 = _____.

4. 12 − 8

12 − _____ = ▢

_____ − _____ = ▢

So, 12 − 8 = _____.

5. 11 − 5

11 − _____ = ▢

_____ − _____ = ▢

So, 11 − 5 = _____.

6. 14 − 7

14 − _____ = ▢

_____ − _____ = ▢

So, 14 − 7 = _____.

Problem Solving

7. Carly has 13 hats. She gives away 8 of them. How many hats does she have left?

_____ hats

Brain Builders

8. There are 18 sea horses swimming in a group. 9 of the sea horses swim away. How many sea horses are still swimming in the group? Draw a picture to show your thinking.

_____ sea horses

Write Math How do you take apart a number to subtract? Explain.

_____

_____

_____

_____

Name _____

My Homework

Homework Helper

eHelp

Need help? connectED.mcgraw-hill.com

You can make a 10 to subtract more easily.

$$14 - 8$$

$$14 - \underline{4} = \boxed{10}$$

$$10 - \underline{4} = \boxed{6}$$

Helpful Hint
Think $14 - 4 = 10$.
So, break apart
8 as 4 and 4.

$14 - 4 = 10$ and
$10 - 4 = 6$

So, $14 - 8 = 6$.

Practice

Take apart the number to make a 10. Then subtract.

1. $17 - 9$

$17 - \underline{\hspace{1.5cm}} = \boxed{}$

$\underline{\hspace{1.5cm}} - \underline{\hspace{1.5cm}} = \boxed{}$

So, $17 - 9 = \underline{\hspace{1.5cm}}$.

2. $13 - 7$

$13 - \underline{\hspace{1.5cm}} = \boxed{}$

$\underline{\hspace{1.5cm}} - \underline{\hspace{1.5cm}} = \boxed{}$

So, $13 - 7 = \underline{\hspace{1.5cm}}$.

Take apart the number to make a 10. Then subtract.

3. $11 - 2$

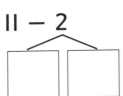

$11 - \underline{\hspace{2cm}} = \boxed{}$

$\underline{\hspace{2cm}} - \underline{\hspace{2cm}} = \boxed{}$

So, $11 - 2 = \underline{\hspace{2cm}}$.

4. $16 - 8$

$16 - \underline{\hspace{2cm}} = \boxed{}$

$\underline{\hspace{2cm}} - \underline{\hspace{2cm}} = \boxed{}$

So, $16 - 8 = \underline{\hspace{2cm}}$.

Brain Builders

Draw a picture to help you solve the problem.

5. There are 18 coconuts on a tree.
   9 coconuts fall off the tree. How
   many coconuts are still on the tree?

$\underline{\hspace{2cm}}$ coconuts

6. **Test Practice** Sadie collects 15 shells at the
   beach. She gives her brother 6 of them.
   How many shells does Sadie have left?

   18 shells   12 shells   11 shells   9 shells

   ○   ○   ○   ○

Name _____

 Math in My World

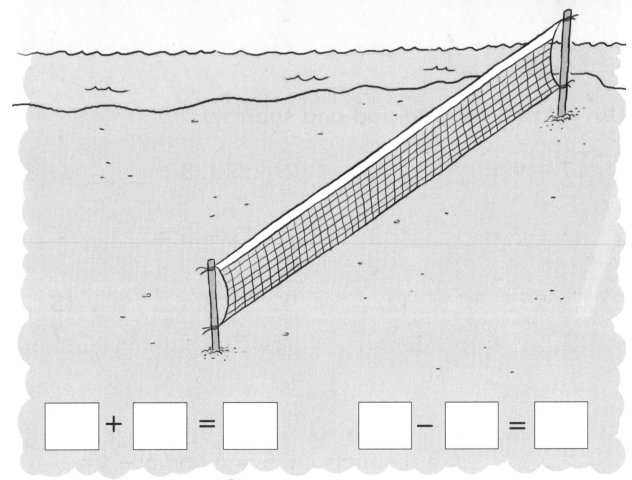

☐ + ☐ = ☐   ☐ − ☐ = ☐

 Teacher Directions: Use 🎲 to model. There are 6 players on one side of the net and 8 players on the other side. How many people are playing volleyball in all? Write the addition number sentence. Then write a related subtraction fact.

Guided Practice

Related facts use the same numbers.
These facts can help you add and subtract.
Find $11 - 5$.

$6 + 5 =$ ____

Helpful Hint
Use $6 + 5 = 11$ to
find $11 - 5 = 6$.

$11 - 5 =$ _6_

Use related facts to add and subtract.

1. $7 + 9 =$ ____

   $16 - 7 =$ ____

2. $5 + 8 =$ ____

   $13 - 5 =$ ____

3.
$$\begin{array}{r} 5 \\ + 7 \\ \hline \end{array}$$
$$\begin{array}{r} 12 \\ - 5 \\ \hline \end{array}$$

4.
$$\begin{array}{r} 8 \\ + 7 \\ \hline \end{array}$$
$$\begin{array}{r} 15 \\ - 7 \\ \hline \end{array}$$

Talk Math Are the facts $1 + 5 = 6$ and $6 - 1 = 5$
related facts? How do you know?

Name ....

Use related facts to add and subtract.

5. 9 + 6 = _____

    15 − 9 = _____

6. 6 + 7 = _____

    13 − 6 = _____

7. 3 + 9 = _____

    12 − 3 = _____

8. 8 + 9 = _____

    17 − 8 = _____

9.   6   14
       + 8   − 6

10.   8   12
       + 4   − 4

Subtract. Write an addition fact to check your subtraction.

11.   16 − 9 = _____

    ____ + ____ = ____

12.   12 − 7 = _____

    ____ + ____ = ____

13.   14 − 9 = _____

    ____ + ____ = ____

14.   11 − 4 = _____

    ____ + ____ = ____

Problem Solving

**Write a subtraction number sentence.
Then write a related addition fact.**

15. Bailey sees 15 birds sitting on a rock.
7 of the birds fly away. How many
of the birds are still on the rock?

_____ – _____ = _____   _____ + _____ = _____

Brain Builders

16. Andre collects 10 shells. He loses 6 of them.
How many shells does Andre have left?

_____ – _____ = _____   _____ + _____ = _____

If Andre loses 2 more shells
how many would he have now?   _____ shells

Write Math   How can related facts help you
add and subtract? Explain.

_____

- -

_____

- -

_____

_____

- -

Name _____

My Homework →

Homework Helper   Need help? ⟋ connectED.mcgraw-hill.com

Related facts can help you add and subtract.

$$8 + 4 = 12 \qquad\qquad 9 + 6 = 15$$
$$12 - 4 = 8 \qquad\qquad 15 - 9 = 6$$

Practice

Use related facts to add and subtract.

1. $5 + 9 =$ _____

    $14 - 5 =$ _____

2. $9 + 5 =$ _____

    $14 - 9 =$ _____

3. $6 + 5 =$ _____

    $11 - 6 =$ _____

4. $9 + 9 =$ _____

    $18 - 9 =$ _____

5. $\begin{array}{r} 8 \\ + 8 \\ \hline \end{array}$ $\begin{array}{r} 16 \\ - 8 \\ \hline \end{array}$

6. $\begin{array}{r} 8 \\ + 5 \\ \hline \end{array}$ $\begin{array}{r} 13 \\ - 8 \\ \hline \end{array}$

Subtract. Write an addition fact to check your subtraction.

7. 15 − 8 = _____

_____ + _____ = _____

8. 17 − 9 = _____

_____ + _____ = _____

Brain Builders

Write a subtraction number sentence. Then write a related addition fact.

9. There are 16 lobsters swimming together. 7 of them swim away. How many lobsters are still swimming together?

_____ − _____ = _____

_____ + _____ = _____

Explain to a family member or friend how you solved the problem.

10. **Test Practice** Mark the related addition fact.

7 − 3 = 4

7 + 3 = 10   7 − 4 = 3   7 + 1 = 8   3 + 4 = 7
○   ○   ○   ○

 Math at Home Write an addition fact such as 3 + 9 = 12. Ask your child to write a related subtraction fact.

Lesson 7
Fact Families

ESSENTIAL QUESTION
What strategies can I use to subtract?

 Math in My World

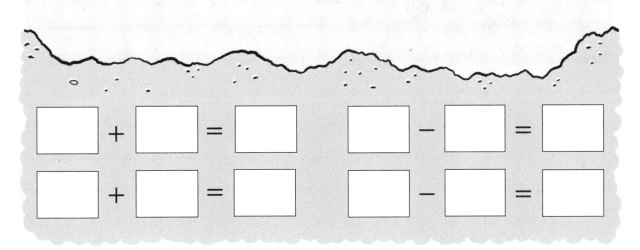

$$\boxed{} + \boxed{} = \boxed{} \qquad \boxed{} - \boxed{} = \boxed{}$$

$$\boxed{} + \boxed{} = \boxed{} \qquad \boxed{} - \boxed{} = \boxed{}$$

 Teacher Directions: Use ▇ to model. There are 7 fish swimming together in a group. 3 more fish join them. Draw a picture to show the story. Write the missing numbers to show the fact family.

Guided Practice

Related facts make a **fact family**.

$5 + 6 = \boxed{11}$ $11 - 5 = \boxed{6}$

$6 + 5 = \boxed{11}$ $11 - 6 = \boxed{5}$

Add and subtract. Complete the fact family.

1.
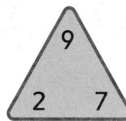

$7 + 2 = \boxed{}$ $9 - 7 = \boxed{}$

$2 + 7 = \boxed{}$ $9 - 2 = \boxed{}$

2.
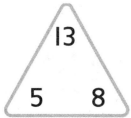

$5 + 8 = \boxed{}$ $13 - 5 = \boxed{}$

$8 + 5 = \boxed{}$ $13 - 8 = \boxed{}$

3.

$6 + 8 = \boxed{}$ $14 - 6 = \boxed{}$

$8 + 6 = \boxed{}$ $14 - 8 = \boxed{}$

Talk Math What fact family can you make with the numbers 4, 9, and 13?

Name ....

Independent Practice

Add and subtract.
Complete the fact family.

In a fact family, all facts have the same numbers.

4.

$9 + 3 =$ [ ] $12 - 9 =$ [ ]

$3 + 9 =$ [ ] $12 - 3 =$ [ ]

5.

$3 + 5 =$ [ ] $8 - 3 =$ [ ]

$5 + 3 =$ [ ] $8 - 5 =$ [ ]

6.
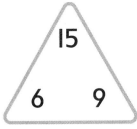

$6 + 9 =$ [ ] $15 - 9 =$ [ ]

$9 + 6 =$ [ ] $15 - 6 =$ [ ]

7.
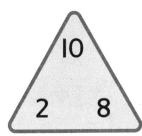

$2 + 8 =$ [ ] $10 - 2 =$ [ ]

$8 + 2 =$ [ ] $10 - 8 =$ [ ]

Problem Solving

8. Joe finds 12 starfish at the beach.
He gives 5 to his grandmother.
How many starfish does he have left?

_____ starfish

Brain Builders

9. There are 16 turtles in the ocean.
7 of those turtles get out. How many
turtles are still in the ocean? Draw
a fact triangle. Include all three
numbers in this problem's fact family.

_____ turtles

10. When I am subtracted from 17, the
difference is 9. What number am I? Explain.

$$17 - \boxed{} = 9$$

_____

_____

_____

_____

_____

Lesson 7

Fact Families

My Homework

Homework Helper

eHelp

Need help? connectED.mcgraw-hill.com

Related facts make up a fact family.

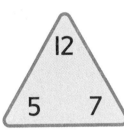

$$5 + 7 = 12 \qquad 12 - 5 = 7$$

$$7 + 5 = 12 \qquad 12 - 7 = 5$$

Practice

Add and subtract. Complete the fact family.

1.

$1 + 9 = \boxed{}$ $\qquad$ $10 - 9 = \boxed{}$

$9 + 1 = \boxed{}$ $\qquad$ $10 - 1 = \boxed{}$

2.

$6 + 9 = \boxed{}$ $\qquad$ $15 - 9 = \boxed{}$

$9 + 6 = \boxed{}$ $\qquad$ $15 - 6 = \boxed{}$

Complete the fact family.

3. Pedro sees 5 flamingos in the water. He sees 4 flamingos on the grass. How many flamingos did he see in all?

☐ + ☐ = ☐   ☐ − ☐ = ☐

☐ + ☐ = ☐   ☐ − ☐ = ☐

Brain Builders

4. Hailey saw 9 crabs on the beach in the morning. She saw 8 more crabs in the afternoon. How many crabs did Hailey see?

☐ + ☐ = ☐   ☐ − ☐ = ☐

☐ + ☐ = ☐   ☐ − ☐ = ☐

Explain how you solved this problem to a family member or friend.

Vocabulary Check

5. Circle the **fact family**.

$2 + 3 = 5$ $4 - 2 = 2$ $7 + 8 = 15$ $15 - 8 = 7$

$3 + 1 = 4$ $5 - 1 = 4$ $8 + 7 = 15$ $15 - 7 = 8$

 Math at Home Challenge your child to write all of the fact families that make 15.

Name ____________________

Lesson 8
Missing Addends

ESSENTIAL QUESTION
What strategies can I use to subtract?

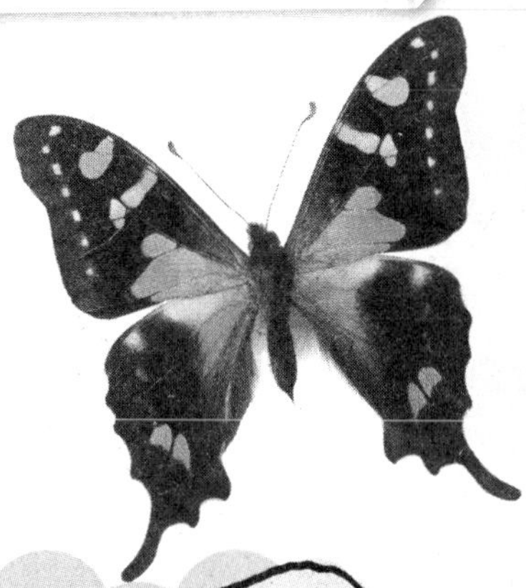

Math in My World

Watch Tools

Part	Part
7	4
Whole	
11	

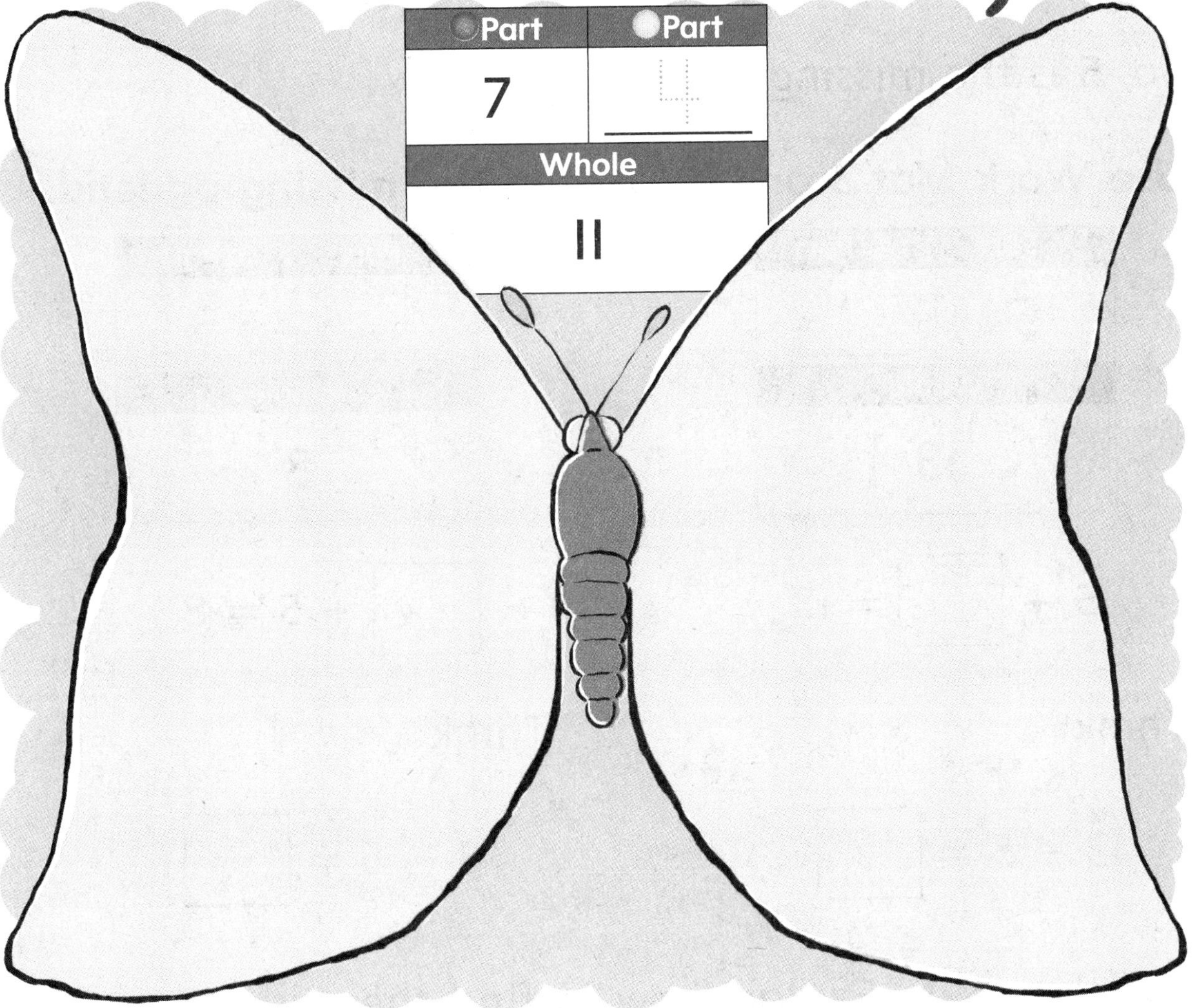

Teacher Directions: Use ●○ to model. The butterfly has 7 spots on the left wing and some more spots on its right wing. It has 11 spots in all. How many spots are on the right wing? Trace your counters to show the spots. Trace the missing addend.

Guided Practice

Use related facts to help you find a **missing addend.**

Helpful Hint
To find the missing part, subtract the known part from the whole.

Part	Part
7	5
Whole	
12	

$7 + \boxed{5} = 12$

$12 - 7 = \boxed{5}$

So, 5 is the missing addend.

Use Work Mat 3 and . Find the missing addend.

1.

Part	Part
5	___
Whole	
13	

$5 + \boxed{} = 13$

Think:

$13 - 5 = \boxed{}$

2.

Part	Part
___	5
Whole	
8	

$\boxed{} + 5 = 8$

Think:

$8 - 5 = \boxed{}$

Talk Math Explain how to find the missing addend in $\boxed{} + 5 = 14$.

Independent Practice

Use Work Mat 3 and ●◯. Find the missing addend.

3.

◯Part	◯Part
4	_____
Whole	
10	

$4 + \boxed{} = 10$

$10 - 4 = \boxed{}$

4.

◯Part	◯Part
_____	6
Whole	
14	

$\boxed{} + 6 = 14$

$14 - 6 = \boxed{}$

5. $8 + \boxed{} = 9$

$9 - 8 = \boxed{}$

6. $\boxed{} + 3 = 11$

$11 - 3 = \boxed{}$

7. $\boxed{} + 7 = 13$

$13 - 7 = \boxed{}$

8. $6 + \boxed{} = 15$

$15 - 6 = \boxed{}$

9. $9 + \boxed{} = 17$

$17 - 9 = \boxed{}$

10. $9 + \boxed{} = 16$

$16 - 9 = \boxed{}$

Problem Solving

11. Max has 5 shovels and some sand pails at the beach. He has 14 shovels and sand pails in all. How many sand pails does he have?

_____ sand pails

Brain Builders

12. When I am added to 7, the sum is the same as $8 + 4$. What number am I?

_____

Write Math   How do you use subtraction to find a missing addend in an addition problem? Explain.

_____

_____

_____

_____

_____

Name _____

My Homework

Homework Helper

Need help? ⟋ connectED.mcgraw-hill.com

You can use related facts to help you find
a missing addend.

●Part	●Part
5	4
Whole	
9	

$5 + \boxed{4} = 9$

$9 - 5 = \boxed{4}$

Helpful Hint
To find the missing part,
subtract the known part
from the whole.

Practice

Find the missing addend.

1.

●Part	●Part
8	___
Whole	
11	

$8 + \boxed{} = 11$

$11 - 8 = \boxed{}$

2.

●Part	●Part
___	8
Whole	
16	

$\boxed{} + 8 = 16$

$16 - 8 = \boxed{}$

Find the missing addend.

3. $9 + \boxed{} = 18$

 $18 - 9 = \boxed{}$

4. $\boxed{} + 6 = 14$

 $14 - 6 = \boxed{}$

5. $\boxed{} + 8 = 15$

 $15 - 8 = \boxed{}$

6. $5 + \boxed{} = 11$

 $11 - 5 = \boxed{}$

 Brain Builders

Write number sentences to solve the problem.

7. There are 16 children flying kites on the beach. Some of the children go home. 9 children are still flying kites. How many children went home?

 ____ = ____ + ____

 ____ = ____ − ____   _____ children

Vocabulary Check

8. Circle the number sentence that shows a **missing addend**.

 $4 + 8 = 12$ $7 + \boxed{} = 15$

 Math at Home Ask your child to tell you the subtraction fact that will help him or her find the missing addend in $8 + \square = 15$.

Fluency Practice

Subtract.

1. 5 − 3 = _____

2. 18 − 9 = _____

3. 4 − 2 = _____

4. 10 − 6 = _____

5. 11 − 5 = _____

6. 7 − 0 = _____

7. 9 − 9 = _____

8. 4 − 1 = _____

9. 14 − 6 = _____

10. 3 − 0 = _____

11. 1 − 1 = _____

12. 8 − 3 = _____

13. 13 − 6 = _____

14. 6 − 3 = _____

15. 5 − 4 = _____

16. 16 − 8 = _____

17. 4 − 4 = _____

18. 15 − 6 = _____

19. 10 − 8 = _____

20. 17 − 8 = _____

21. 9 − 1 = _____

22. 7 − 3 = _____

23. 14 − 7 = _____

24. 10 − 5 = _____

Fluency Practice

Subtract.

1. $\begin{array}{r} 10 \\ -\ 3 \\ \hline \end{array}$

2. $\begin{array}{r} 7 \\ -\ 6 \\ \hline \end{array}$

3. $\begin{array}{r} 14 \\ -\ 7 \\ \hline \end{array}$

4. $\begin{array}{r} 8 \\ -\ 4 \\ \hline \end{array}$

5. $\begin{array}{r} 2 \\ -\ 0 \\ \hline \end{array}$

6. $\begin{array}{r} 17 \\ -\ 9 \\ \hline \end{array}$

7. $\begin{array}{r} 10 \\ -\ 5 \\ \hline \end{array}$

8. $\begin{array}{r} 5 \\ -\ 5 \\ \hline \end{array}$

9. $\begin{array}{r} 6 \\ -\ 1 \\ \hline \end{array}$

10. $\begin{array}{r} 11 \\ -\ 3 \\ \hline \end{array}$

11. $\begin{array}{r} 4 \\ -\ 2 \\ \hline \end{array}$

12. $\begin{array}{r} 12 \\ -\ 6 \\ \hline \end{array}$

13. $\begin{array}{r} 1 \\ -\ 0 \\ \hline \end{array}$

14. $\begin{array}{r} 3 \\ -\ 2 \\ \hline \end{array}$

15. $\begin{array}{r} 8 \\ -\ 4 \\ \hline \end{array}$

16. $\begin{array}{r} 15 \\ -\ 7 \\ \hline \end{array}$

17. $\begin{array}{r} 13 \\ -\ 5 \\ \hline \end{array}$

18. $\begin{array}{r} 8 \\ -\ 8 \\ \hline \end{array}$

19. $\begin{array}{r} 9 \\ -\ 3 \\ \hline \end{array}$

20. $\begin{array}{r} 10 \\ -\ 1 \\ \hline \end{array}$

21. $\begin{array}{r} 6 \\ -\ 2 \\ \hline \end{array}$

22. $\begin{array}{r} 3 \\ -\ 3 \\ \hline \end{array}$

23. $\begin{array}{r} 16 \\ -\ 9 \\ \hline \end{array}$

24. $\begin{array}{r} 11 \\ -\ 2 \\ \hline \end{array}$

Name ....

Vocabulary Check

Circle the correct answer.

1. count back

6 − 2 = _____

0 1 2 3 4 5 6 7 8 9 10

6 − 2 = _____

0 1 2 3 4 5 6 7 8 9 10

2. fact family

7 + 3 = 10
3 + 7 = 10
10 − 7 = 3
10 − 3 = 7

6 + 4 = 10
4 + 6 = 10
5 + 5 = 10
8 + 2 = 10

3. missing addend

8 + 3 = 11

8 + ☐ = 11

4. difference

↓
9 − 4 = 5

↓
6 − 1 = 5

Concept Check

Count back to subtract.

5. 7 − 3 = _____

6. 9 − 2 = _____

Use the number line to help you subtract.

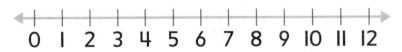

$$0 \quad 1 \quad 2 \quad 3 \quad 4 \quad 5 \quad 6 \quad 7 \quad 8 \quad 9 \quad 10 \quad 11 \quad 12$$

7. 11 − 2 = _____

8. 10 − 1 = _____

Add the doubles facts. Then subtract.

9. 6 + 6 = _____

12 − 6 = _____

10. 8 + 8 = _____

16 − 8 = _____

Add and subtract. Complete the fact family.

11.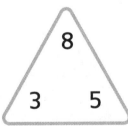

3 + 5 = ☐   8 − 3 = ☐

5 + 3 = ☐   8 − 5 = ☐

Find the missing addend.

12. 5 + ☐ = 12

12 − 5 = ☐

13. 9 + ☐ = 14

14 − 9 = ☐

Problem Solving

14. Paige writes two related facts using these numbers. What related facts could she have written?

12, 8, 4

_____ + _____ = _____

_____ − _____ = _____

Brain Builders

15. Jayson caught 15 fish. Jayson caught 8 more fish than Ashlyn. How many fish did Ashlyn catch?

_____ − _____ = _____ fish

16. **Test Practice** Cris has 14 jump ropes. What doubles fact shows the number of jump ropes Cris has?

6 + 8 = 14 jump ropes ○   7 + 7 = 14 jump ropes ○

14 − 5 = 9 jump ropes ○   6 + 6 = 12 jump ropes ○

Show the ways to subtract.

ESSENTIAL QUESTION

What strategies can I use to subtract?

Count Back

12, _____, _____, _____

12 − 3 = _____

Use Doubles

10 + 10 = _____

20 − 10 = _____

Make a Ten

17 − 8

_____ − _____ = ☐

_____ − _____ = ☐

So, 17 − 8 = _____.

Missing Addend

5 + ☐ = 14

14 − 5 = ☐

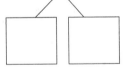

Now I Know!

Name _____

Date _____

Score _____

Performance Task

On a Hike

Kawan and his family were taking a hike.
They saw many things.

Show all of your work to receive full credit.

Part A

Kawan saw 14 white and yellow flowers. 5 flowers were white. Use the number line to find how many flowers were yellow. Write the number.

_____ yellow flowers

Part B

Kawan picked 17 apples. He ate 8 apples. How many apples did he have left? Write a subtraction number sentence. Then write a related fact.

_____ − _____ = _____

_____ + _____ = _____

Part C

Kawan saw 12 ducks swimming in a pond. 5 ducks flew out of the pond. How many ducks were left? Take apart the number to make a 10. Then subtract.

$$12 - 5$$

$12 - 5 = $ _____ ducks left

Part D

Kawan found 14 feathers. He gave 5 feathers to his brother. How many feathers does he have left? Complete the fact family.

_____ + _____ = _____

_____ + _____ = _____

_____ − _____ = _____

_____ − _____ = _____

Glossary/Glosario

Aa	English	Spanish/Español

add (adding, addition) To join together sets to find the total or sum.

$$2 + 5 = 7$$

sumar (adición) Unir conjuntos para hallar el total o la suma.

$$2 + 5 = 7$$

addend Any numbers or quantities being added together.

$$2 + 3$$

2 is an addend and 3 is an addend.

sumando Números o cantidades que se suman.

$$2 + 3$$

2 es un sumando y 3 es un sumando.

addition number sentence An expression using numbers and the + and = signs.

$$4 + 5 = 9$$

enunciado numérico de suma Expresión en la cual se usan números con los signos + e =.

$$4 + 5 = 9$$

Aa

after To follow in place or time.

6 is just *after* 5

después Que sigue en lugar o en tiempo.

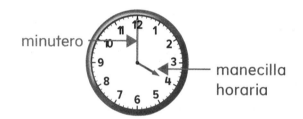

6 está justo *después* de 5.

analog clock A clock that has an hour hand and a minute hand.

minute hand → / ← hour hand

reloj analógico Reloj que tiene manecilla horaria y minutero.

minutero → / ← manecilla horaria

Bb

bar graph A graph that uses bars to show data.

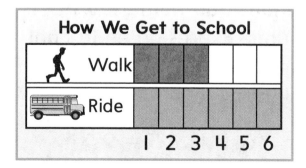

How We Get to School

		1	2	3	4	5	6
🚶	Walk						
🚌	Ride						

gráfica de barras Gráfica que usa barras para ilustrar datos.

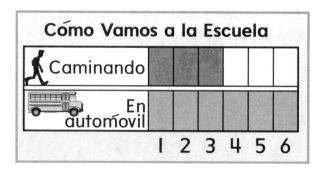

Cómo Vamos a la Escuela

		1	2	3	4	5	6
🚶	Caminando						
🚌	En automóvil						

before

5 6 7 8

6 is just *before* 7.

antes

5 6 7 8

6 está justo *antes* del 7.

between

The kitten is *between* the two dogs.

entre

El gatito está *entre* dos perros.

Cc

capacity The amount of dry or liquid material a container can hold.

capacidad Cantidad de material seco o líquido que cabe en un recipiente.

Cc

cent ¢

1¢   1 cent

centavo ¢

1¢   1 centavo

circle A closed round shape.

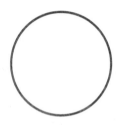

círculo Figura redonda y cerrada.

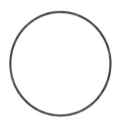

compare Look at objects, shapes, or numbers and see how they are alike or different.

comparar Observar objetos, formas o números para saber en qué se parecen y en qué se diferencian.

cone A three-dimensional shape that narrows to a point from a circular face.

cono Una figura tridimensional que se estrecha hasta un punto desde una cara circular.

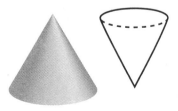

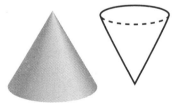

count back On a number line, start at the number 5 and count back 3.

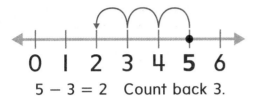

5 − 3 = 2   Count back 3.

contar hacia atrás En una recta numérica, comienza en el número 5 y cuenta 3 hacia atrás.

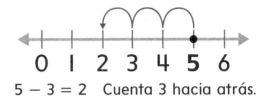

5 − 3 = 2   Cuenta 3 hacia atrás.

count on (or count up) On a number line, start at the number 4 and count up 2.

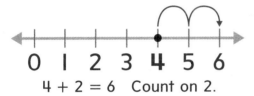

4 + 2 = 6   Count on 2.

seguir contando (o contar hacia delante) En una recta numérica, comienza en el 4 y cuenta 2.

4 + 2 = 6   Cuenta 2 hacia delante.

cube A three-dimensional shape with 6 square faces.

cubo Una figura tridimensional con 6 caras cuadradas.

cylinder A three-dimensional shape that is shaped like a can.

cilindro Una figura tridimensional que tiene la forma de una lata.

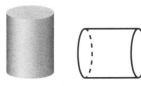

data Numbers or symbols collected to show information.

datos Números o símbolos que se recopilan para mostrar información.

Name	Number of Pets
Maria	3
James	1
Alonzo	4

Nombre	Número de mascotas
Maria	3
James	1
Alonzo	4

day

day

día

día

difference The answer to a subtraction problem.

$$3 - 1 = 2$$

The difference is 2.

diferencia Resultado de un problema de resta.

$$3 - 1 = 2$$

La diferencia es 2.

digital clock A clock that uses only numbers to show time.

reloj digital Reloj que usa solo números para mostrar la hora.

dime 10¢ or 10 cents

head   tail

moneda de 10¢ 10¢ o 10 centavos

cara   cruz

doubles (doubles plus 1, near doubles) Two addends that are the same number.

$2 + 2 = 4$

$2 + 3 = 5$ $2 + 1 = 3$

dobles (y dobles más 1, casi dobles) Dos sumandos que son el mismo número.

$2 + 2 = 4$

$2 + 3 = 5$ $2 + 1 = 3$

equal parts Each part is the same size.

A muffin cut in equal parts.

partes iguales Cada parte es del mismo tamaño.

Un panecillo cortado en partes iguales.

equal to =

6 = 6
6 is equal to 6.

igual a =

6 = 6
6 es igual a 6.

equals (=) Having the same value as or is the same as.

2 + 4 = 6

equals sign ↑

igual (=) Que tienen el mismo valor o son lo mismo.

2 + 4 = 6

signo igual ↑

face The flat part of a three-dimensional shape.

cara La parte plana de una figura tridimensional.

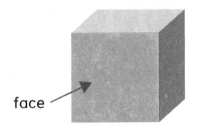

face

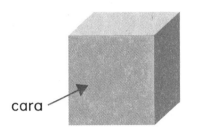

cara

fact family Addition and subtraction sentences that use the same numbers. Sometimes called *related facts*.

familia de operaciones Enunciados de suma y resta que tienen los mismos números. Algunas veces se llaman *operaciones relacionadas*.

$6 + 7 = 13$ $13 - 7 = 6$

$7 + 6 = 13$ $13 - 6 = 7$

$6 + 7 = 13$ $13 - 7 = 6$

$7 + 6 = 13$ $13 - 6 = 7$

false Something that is not a fact. The opposite of true.

falso Algo que no es cierto. Lo opuesto de verdadero.

Ff

fewer/fewest The number or group with less.

There are fewer yellow counters than red ones.

menos/el menor El número o grupo con menos.

Hay menos fichas amarillas que fichas rojas.

fourths Four equal parts of a whole. Each part is a fourth, or a quarter of the whole.

cuartos Cuatro partes iguales de un todo. Cada parte es un cuarto, o la cuarta parte del todo.

graph A way to present data collected.

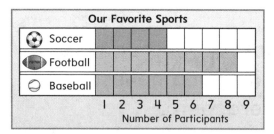

bar graph

gráfica Forma de presentar datos recopilados.

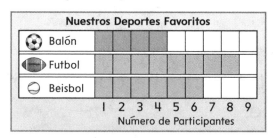

gráfica de barras

greater than (>)/greatest The number or group with more.

| 4 | 23 | 56 |

56 is the greatest.

mayor que (>)/el mayor El número o grupo con más cantidad.

| 4 | 23 | 56 |

56 es el mayor.

half hour (or half past)
One half of an hour is 30 minutes. Sometimes called *half past* or *half past the hour*.

halves Two equal parts of a whole. Each part is a half of the whole.

heavy (heavier, heaviest)
Weighs more.

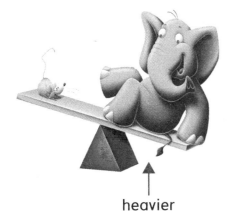

heavier

An elephant is heavier than a mouse.

media hora (o y media)
Media hora son 30 minutos. A veces se dice *hora y media*.

mitades Dos partes iguales de un todo. Cada parte es la mitad de un todo.

pesado (más pesado, el más pesado) Pesa más.

más pesado

Un elefante es más pesado (pesa más) que un ratón.

Hh

height

short tall

altura

bajo alto

hexagon A two-dimensional shape that has six sides.

hexágono Figura bidimensional que tiene seis lados.

holds less/least

The glass holds less than the pitcher.

contener menos

El vaso contiene menos que la jarra.

holds more/most

The pitcher holds more than the glass.

hour A unit of time.

I hour = 60 minutes

hour hand The hand on a clock that tells the hour. It is the shorter hand.

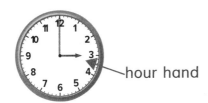

hour hand

contener más

La jarra contiene más que el vaso.

hora Unidad de tiempo.

I hora = 60 minutos

manecilla horaria
Manecilla del reloj que indica la hora. Es la manecilla más corta.

manecilla horaria

Hh

hundreds The numbers in the range of 100-999. It is the place value of a number.

centenas Los números en el rango del 100 al 999. Es el valor posicional de un número.

Ii

inverse Operations that undo each other.

Addition and subtraction are inverse or opposite operations.

operaciónes inversas Operaciones que se anulan entre sí.

La suma y la resta son operaciones inversas u opuestas.

Ll

length

length

longitud

longitud

less than (<)/least The number or group with fewer.

| 4 | 23 | 56 |

4 is the least.

menor que (<)/el menor El número o grupo con menos cantidad.

| 4 | 23 | 56 |

4 es el menor.

light (lighter, lightest) Weighs less.

lighter

The mouse is lighter than the elephant.

liviano (más liviano, el más liviano) Pesa menos.

más liviano

El ratón es más liviano (pesa menos) que el elefante.

long (longer, longest) A way to compare the lengths of two objects.

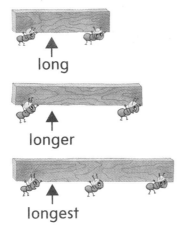

long

longer

longest

largo (más largo, el más largo) Forma de comparar la longitud de dos objetos.

largo

más largo

el más largo

mass The amount of matter in an object. The mass of an object never changes.

masa Cantidad de materia en un objeto. La masa de un cuerpo nunca cambia.

measure To find the length, height, weight or capacity using standard or nonstandard units.

medir Hallar la longitud, altura, peso o capacidad mediante unidades estándar o no estándar.

minus (−) The sign used to show subtraction.

menos (−) Signo que indica resta.

$$5 - 2 = 3$$

minus sign

$$5 - 2 = 3$$

signo menos

minute (min) A unit to measure time.

minuto (min) Unidad que se usa para medir el tiempo.

I minute = 60 seconds

I minuto = 60 segundos

minute hand The longer hand on a clock that tells the minutes.

minute hand

minutero La manecilla más larga del reloj. Indica los minutos.

minutero

missing addend

$$9 + \underline{\qquad} = 16$$

The missing addend is 7.

sumando desconocido

$$9 + \underline{\qquad} = 16$$

El sumando desconocido es 7.

month

month

mes

mes

Mm

more

← more

más

← más

nickel 5¢ or
5 cents

head   tail

moneda de 5¢ 5¢ o
5 centavos

cara   cruz

number Tells how many.
1, 2, 3, 4, 5, 6, 7, 8, 9, 10 ...

There are 3 chicks.

número Dice cuántos hay.
1, 2, 3, 4, 5, 6, 7, 8, 9, 10 ...

Hay tres pollitos.

number line A line with
number labels.

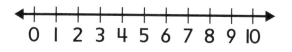

recta numérica Recta con
marcas de números.

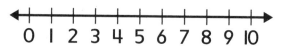

o'clock At the beginning of the hour.

It is 3 o'clock.

en punto Al comienzo de la hora.

Son las 3 en punto.

ones The numbers in the range of 0–9. It is the place value of a number.

unidades Los números en el rango de 0 a 9. Es el valor posicional de un número.

order

1, 3, 6, 7, 9

These numbers are in order from least to greatest.

orden

1, 3, 6, 7, 9

Estos números están en orden del menor al mayor.

ordinal number

first   second   third

númeral ordinal

primero segundo tercero

part One of the parts joined when adding.

Part	Part
2	2
Whole	

parte Una de las partes que se juntan al sumar.

Parte	Parte
2	2
El total	

pattern An order that a set of objects or numbers follows over and over.

A, A, B, A, A, B, A, A, B

—pattern unit

patrón Orden que sigue continuamente un conjunto de objetos o números.

A, A, B, A, A, B, A, A, B

—unidad de patrón

penny I¢ or I cent

head　　　tail

moneda de I¢ I¢ o I centavo

cara　　　cruz

picture graph A graph that has different pictures to show information collected.

gráfica con imágenes Gráfica que tiene diferentes imágenes para ilustrar la información recopilada.

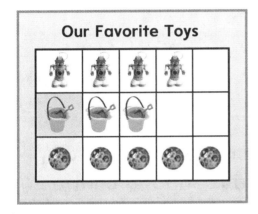

place value The value given to a digit by its place in a number.

53

5 is in the tens place.
3 is in the ones place.

valor posicional Valor de un *dígito* según el lugar en el número.

53

5 está en el lugar de las decenas.
3 está en el lugar de las unidades.

plus (+) The sign used to show addition.

$$4 + 5 = 9$$
↑
plus sign

más (+) Símbolo para mostrar la suma.

$$4 + 5 = 9$$
↑
signo más

Pp

position Tells where an object is.

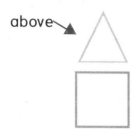

posición Indica dónde está un objeto.

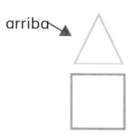

Rr

rectangle A shape with four sides and four corners.

rectángulo Figura con cuatro lados y cuatro esquinas.

rectangular prism A three-dimensional shape with 6 faces that are rectangles.

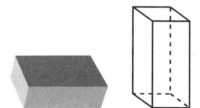

prisma rectangular Una figura tridimensional con 6 caras que son rectángulos.

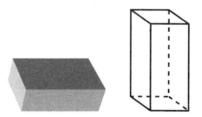

regroup To take apart a number and write it in a new way.

1 ten + 2 ones becomes 12 ones.

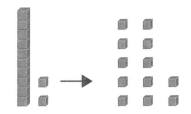

reagrupar Separar un número para escribirlo en una nueva forma.

1 decena + 2 unidades se convierten en 12 unidades.

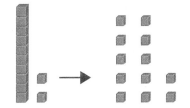

related fact(s) Basic facts using the same numbers. Sometimes called a *fact family*.

$$4 + 1 = 5 \qquad 5 - 4 = 1$$
$$1 + 4 = 5 \qquad 5 - 1 = 4$$

operaciones relacionadas Operaciones básicas en las cuales se usan los mismos números. También se llaman *familias de operaciones*.

$$4 + 1 = 5 \qquad 5 - 4 = 1$$
$$1 + 4 = 5 \qquad 5 - 1 = 4$$

repeating pattern

patrón repetitivo

short (shorter, shortest) To compare length or height of two (or more) objects.

corto (más corto, el más corto) Comparar la longitud o la altura de dos (o más) objetos.

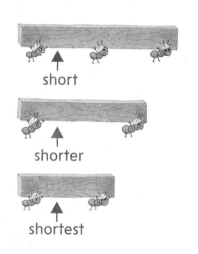

short

shorter

shortest

corto

más corto

el más corto

side

lado

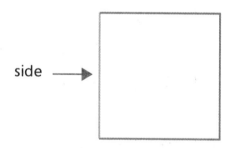

side →

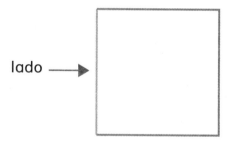

lado →

sort To group together like items.

clasificar Agrupar elementos con iguales características.

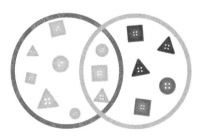

sphere A solid shape that has the shape of a round ball.

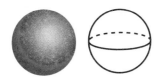

esfera Un sólido con la forma de una pelota redonda.

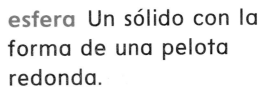

square A rectangle that has four equal sides.

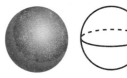

cuadrado Rectángulo que tiene cuatro lados iguales.

subtract (subtracting, subtraction) To take away, take apart, separate, or find the difference between two sets. The opposite of addition.

$$4 - 1 = 3$$

restar (resta, sustracción) Eliminar, quitar, separar o hallar la diferencia entre dos conjuntos. Lo opuesto de la suma.

$$4 - 1 = 3$$

subtraction number sentence An expression using numbers and the − and = signs.

$$9 - 5 = 4$$

enunciado numérico de resta Expresión en la cual se usan números con los signos − e =.

$$9 - 5 = 4$$

sum The answer to an addition problem.

$$2 + 4 = \underset{\uparrow}{6}$$
sum

suma Resultado de la operación de sumar.

$$2 + 4 = \underset{\uparrow}{6}$$
suma

survey To collect data by asking people the same question.

encuesta Recopilación de datos haciendo las mismas preguntas a un grupo de personas.

Favorite Foods	
Food	Votes
🍎	卌
🌽	|||
🌮	卌 |||

This survey shows favorite foods.

Comidas favoritas	
Comida	Votos
🍎	卌
🌽	|||
🌮	卌 |||

Esta encuesta muestra las comidas favoritas.

tall (taller, tallest)

tall

alto (más alto, el más alto)

alto

tally chart A way to show data collected using tally marks.

Favorite Foods	
Food	Votes
🍎	ⵜⵜⵜ
🌽	|||
🥪	ⵜⵜⵜ |||

tabla de conteo Forma de mostrar los datos recopilados utilizando marcas de conteo.

Comidas favoritas	
Comida	Votos
🍎	ⵜⵜⵜ
🌽	|||
🥪	ⵜⵜⵜ |||

tens The numbers in the range 10–99. It is the place value of a number.

53

5 is in the tens place.
3 is in the ones place.

decenas Los números en el rango del 10 al 99. Es el valor posicional de un número.

53

5 está en el lugar de las decenas.
3 está en el lugar de las unidades.

three-dimensional shape
A solid shape.

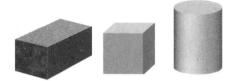

figura tridimensional
Un sólido.

trapezoid A four-sided plane shape with only two opposite sides that are parallel.

trapecio Figura de cuatro lados con solo dos lados opuestos que son paralelos.

triangle A shape with three sides.

triángulo Figura con tres lados.

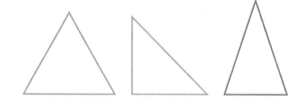

true Something that is a fact. The opposite of false.

verdadero Algo que es cierto. Lo opuesto de falso.

two-dimensional shape
The outline of a shape
such as a triangle, square,
or rectangle.

figura bidimensional
Contorno de una figura
como un triángulo,
o un cuadrado rectángulo.

Uu

unit An object used to
measure.

unidad Objeto que se usa
para medir.

Vv

Venn diagram A drawing
that uses circles to sort
and show data.

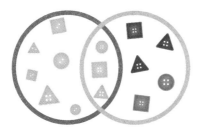

diagrama de Venn Dibujo
que tiene círculos para
clasificar y mostrar datos.

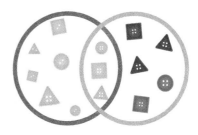

Vv

vertex

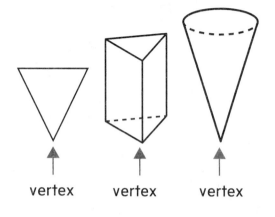

vertex   vertex   vertex

vértice

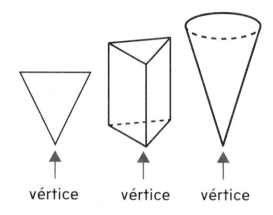

vértice   vértice   vértice

weight

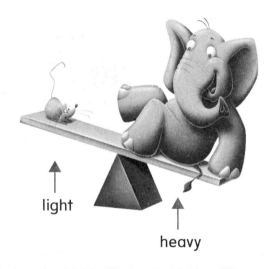

light

heavy

peso

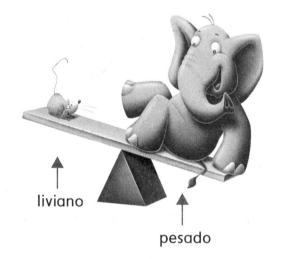

liviano

pesado

whole The entire amount of an object.

el todo La cantidad total o el objeto completo.

year

año

January						
S	M	T	W	T	F	S
						1
2	3	4	5	6	7	8
9	10	11	12	13	14	15
16	17	18	19	20	21	22
23	24	25	26	27	28	29
30	31					

February						
S	M	T	W	T	F	S
		1	2	3	4	5
6	7	8	9	10	11	12
13	14	15	16	17	18	19
20	21	22	23	24	25	26
27	28					

enero						
d	l	m	m	j	v	s
						1
2	3	4	5	6	7	8
9	10	11	12	13	14	15
16	17	18	19	20	21	22
23	24	25	26	27	28	29
30	31					

febrero						
d	l	m	m	j	v	s
		1	2	3	4	5
6	7	8	9	10	11	12
13	14	15	16	17	18	19
20	21	22	23	24	25	26
27	28					

March						
S	M	T	W	T	F	S
		1	2	3	4	5
6	7	8	9	10	11	12
13	14	15	16	17	18	19
20	21	22	23	24	25	26
27	28	29	30	31		

April						
S	M	T	W	T	F	S
					1	2
3	4	5	6	7	8	9
10	11	12	13	14	15	16
17	18	19	20	21	22	23
24	25	26	27	28	29	30

marzo						
d	l	m	m	j	v	s
		1	2	3	4	5
6	7	8	9	10	11	12
13	14	15	16	17	18	19
20	21	22	23	24	25	26
27	28	29	30	31		

abril						
d	l	m	m	j	v	s
					1	2
3	4	5	6	7	8	9
10	11	12	13	14	15	16
17	18	19	20	21	22	23
24	25	26	27	28	29	30

May						
S	M	T	W	T	F	S
1	2	3	4	5	6	7
8	9	10	11	12	13	14
15	16	17	18	19	20	21
22	23	24	25	26	27	28
29	30	31				

June						
S	M	T	W	T	F	S
			1	2	3	4
5	6	7	8	9	10	11
12	13	14	15	16	17	18
19	20	21	22	23	24	25
26	27	28	29	30		

mayo						
d	l	m	m	j	v	s
1	2	3	4	5	6	7
8	9	10	11	12	13	14
15	16	17	18	19	20	21
22	23	24	25	26	27	28
29	30	31				

junio						
d	l	m	m	j	v	s
			1	2	3	4
5	6	7	8	9	10	11
12	13	14	15	16	17	18
19	20	21	22	23	24	25
26	27	28	29	30		

July						
S	M	T	W	T	F	S
					1	2
3	4	5	6	7	8	9
10	11	12	13	14	15	16
17	18	19	20	21	22	23
24	25	26	27	28	29	30
31						

August						
S	M	T	W	T	F	S
	1	2	3	4	5	6
7	8	9	10	11	12	13
14	15	16	17	18	19	20
21	22	23	24	25	26	27
28	29	30	31			

julio						
d	l	m	m	j	v	s
					1	2
3	4	5	6	7	8	9
10	11	12	13	14	15	16
17	18	19	20	21	22	23
24	25	26	27	28	29	30
31						

agosto						
d	l	m	m	j	v	s
	1	2	3	4	5	6
7	8	9	10	11	12	13
14	15	16	17	18	19	20
21	22	23	24	25	26	27
28	29	30	31			

September						
S	M	T	W	T	F	S
				1	2	3
4	5	6	7	8	9	10
11	12	13	14	15	16	17
18	19	20	21	22	23	24
25	26	27	28	29	30	

October						
S	M	T	W	T	F	S
						1
2	3	4	5	6	7	8
9	10	11	12	13	14	15
16	17	18	19	20	21	22
23	24	25	26	27	28	29
30	31					

septiembre						
d	l	m	m	j	v	s
				1	2	3
4	5	6	7	8	9	10
11	12	13	14	15	16	17
18	19	20	21	22	23	24
25	26	27	28	29	30	

octubre						
d	l	m	m	j	v	s
						1
2	3	4	5	6	7	8
9	10	11	12	13	14	15
16	17	18	19	20	21	22
23	24	25	26	27	28	29
30	31					

November						
S	M	T	W	T	F	S
		1	2	3	4	5
6	7	8	9	10	11	12
13	14	15	16	17	18	19
20	21	22	23	24	25	26
27	28	29	30			

December						
S	M	T	W	T	F	S
				1	2	3
4	5	6	7	8	9	10
11	12	13	14	15	16	17
18	19	20	21	22	23	24
25	26	27	28	29	30	31

noviembre						
d	l	m	m	j	v	s
		1	2	3	4	5
6	7	8	9	10	11	12
13	14	15	16	17	18	19
20	21	22	23	24	25	26
27	28	29	30			

diciembre						
d	l	m	m	j	v	s
				1	2	3
4	5	6	7	8	9	10
11	12	13	14	15	16	17
18	19	20	21	22	23	24
25	26	27	28	29	30	31

zero The number zero equals none or nothing.

cero El número cero es igual a nada o ninguno.

Name

Work Mat I: Ten-Frame

Work Mat I: Ten-Frame   WMI

Work Mat 2: Ten-Frames

Work Mat 2: Ten-Frames

Name _____

Part

Part

Whole

Work Mat 4: Number Lines

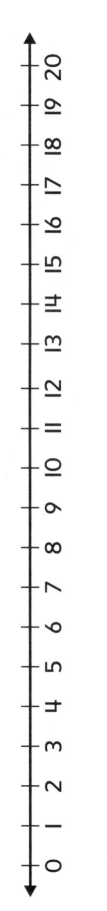

Name

Work Mat 5: Number Lines

61 62 63 64 65 66 67 68 69 70 71 72 73 74 75 76 77 78 79 80

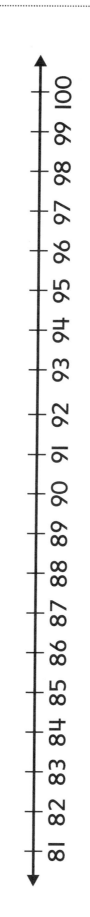

81 82 83 84 85 86 87 88 89 90 91 92 93 94 95 96 97 98 99 100

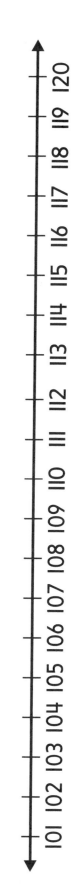

101 102 103 104 105 106 107 108 109 110 111 112 113 114 115 116 117 118 119 120

Work Mat 6: Grid

Work Mat 6: Grid

Name

Work Mat 7: Tens and Ones Chart

Tens	Ones

Work Mat 8: Hundreds, Tens, and Ones Chart

Hundreds	Tens	Ones

Work Mat 8: Hundreds, Tens, and Ones Chart